Bringing Nature Home

"A fascinating study of the trees, shrubs, and vines that feed the insects, birds, and other animals in the suburban garden."
—*New York Times*

"Provides the rationale behind the use of native plants, a concept that has rapidly been gaining momentum... The text makes a case for native plants and animals in a compelling and complete fashion."
—*Washington Post*

"This is the 'it' book in certain gardening circles. It's really struck a nerve."
—*Philadelphia Inquirer*

"Reading this book will give you a new appreciation of the natural world—and how much wild creatures need gardens that mimic the disappearing wild."
—*Minneapolis Star Tribune*

"A compelling argument for the use of native plants in gardens and landscapes."
—*Landscape Architecture*

"An essential guide for anyone interested in increasing biodiversity in the garden."
—*American Gardener*

"Delivers an important message for all gardeners: Choosing native plants fortifies birds and other wildlife and protects them from extinction."
—*WildBird Magazine*

"An informative and engaging account of the ecological interactions between plants and wildlife, this fascinating handbook explains why exotic plants can hinder and confuse native creatures, from birds and bees to larger fauna."
—*Seattle Post-Intelligencer*

"Tallamy explains eloquently how native plant species depend on native wildlife."
—*San Luis Obispo Tribune*

"Will persuade all of us to take a look at what is in our own yards with an eye to how we, too, can make a difference. It has already changed me."
—*Traverse City Record-Eagle*

"There's an increasing interest among homeowners and others to include more native species in their landscape, thanks to books like *Bringing Nature Home,* by Doug Tallamy, which extol the virtues of native plants over exotic ornamentals for attracting and sustaining beneficial insects."
—*Andover Townsman*

"Doug Tallamy weaves an interesting story of how exotic invasive plants affect birds and other components of a healthy forests. It's a compelling and important story to understand."
—*Bradford Era*

THE NATURE OF OAKS

The
Nature
of
Oaks

The Rich Ecology *of* Our Most Essential Native Trees

DOUGLAS W. TALLAMY

Timber Press · Portland, Oregon

Photos on pages 104 by Sara Bright, 109 by Dave Funk, and 119 by Guy Sternberg. All other photos are by the author.

Published in 2021 by Timber Press, Inc.
The Haseltine Building
133 S.W. Second Avenue, Suite 450
Portland, Oregon 97204-3527
timberpress.com

Printed in China
Text and cover design by Adrianna Sutton

ISBN 978-1-64326-044-0

Catalog records for this book are available from the Library of Congress and the British Library.

FSC
www.fsc.org
MIX
Paper from
responsible sources
FSC® C144853

Contents

Prologue • 9

October • 13
November • 21
December • 27
January • 31
February • 44
March • 51
April • 59
May • 71
June • 88
July • 102
August • 121
September • 142

Epilogue • 152

Acknowledgments • 157
References • 159
How to plant an oak • 163
Best oak options for your area • 167
North American native oaks • 179
Index • 185

Prologue

ON 15 JULY 2000, Cindy and I moved into our newly constructed house in southeastern Pennsylvania. I remember the date well because 15 July is my son's birthday, and one of his birthday presents that year was to help us carry furniture from our rental van into our new house. On one of many trips, a paper wasp stung him on the back of his head just as he passed through the front door. The wasp nest was in the upper corner of the jamb, and its occupants gave my son a painful reminder that he is the tallest Tallamy on record.

Our property is 10 acres, and it had been mowed for hay for decades before we bought it. Its only trees were three hickories in the upper corner of the property, a few black cherries, two black oaks, and an occasional black walnut along an old fence line from the days the area was grazed by cattle. I was eager to plant more trees, so that fall I collected some white oak acorns and planted one in a small flower pot. I don't remember where I found that acorn, but it was probably from a large white oak that grew near the turn-around spot on the jogging/walking route Cindy and I now take when the spirit moves us. Years later we would collect bags of acorns from that tree each fall, mostly because we couldn't stand to see them crunched by cars or mowed by the homeowner.

White oak acorns germinate soon after they fall to the ground, sending what will become the tree's taproot deep into the soil before the first frost.

They then rest for the winter and do not produce the first green shoot until the following spring. Aboveground growth is slow that first year; only one set of true leaves is produced, and the tree remains just a few inches tall. This is a snail's pace compared to trees like American elms or sycamores, which can grow more than 2 feet in their first year, and it probably explains the general perception that oaks are slow growers.

But their growth rate below ground is a different story. All the energy those first leaves gather from the sun is allocated toward root growth. In fact, after their first year, a seedling oak may have up to 10 times more root mass than the biomass of leaves and shoots above the ground.

Oaks produce enormous root systems over their lifetimes, and these help make them champions when it comes to soil stabilization, carbon sequestration, and watershed management. William Bryant Logan (2005) relates the story of a group of Scandinavians who were interested in measuring the extent of the root system of *Quercus robur*, commonly known as the English oak or king's oak, since various kings of England protected forests of these trees for their private hunting grounds. Their approach was not high tech; they simply started scraping the dirt away to see how far the roots extended from the trunk of the tree. After days of digging they had exposed a root system that extended three times the width of the oak's canopy... and then they gave up, convinced that the roots went on forever. I'm sure they did not, but the Scandinavians' efforts certainly obliterated the myth that oak tree roots go out only as far as the crown spread.

I planted my white oak in a flower pot because if I had planted straight into the ground, the chances that a vole or white-footed mouse would have dined on it sometime over the winter was at least 99%. It's even a challenge to keep mice out of flower pots over the winter, but somehow I managed and was able to plant my tiny tree the following June. Oak seedlings are a favorite of browsing white-tailed deer, so I put a small chicken-wire cage around my seedling that first year and then upgraded to a 5-foot galvanized wire cage that kept my oak safe and sound for the next four years. I'm sure I watered my transplant once or twice but, through serendipitous laziness, I did not fertilize the little guy. Although I didn't know it at the time, most North American trees do not need (and many, in fact, do not tolerate) high-nitrogen fertilizers. They are adapted to grow best on the nutrient-poor soils that the last glaciation left behind in North America. This is particularly true for white oaks, which thrive on poor, shallow soils.

It is now 18 years later, and that tiny seedling is nearly 45 feet tall with a trunk circumference of 47 inches at breast height and a canopy spread of 30 feet. Despite this impressive growth rate, my oak is still a baby; oaks in general are long-lived trees, and members of the white oak group, as well as species in the red oak and canyon oak groups, can easily live many hundreds of years if their roots are given free rein and not impeded by sewer lines, roadcuts, basements, parking lots, or some other such human constraint. During that impressive life span a single tree will drop up to 3 million acorns and serve as a lifeline for countless creatures, including dozens of bird species, rodents, bears, raccoons, opossums, rat snakes, fence lizards, several butterflies, hundreds of moths, cynipid gall wasps and other predators and parasitoids, weevils, myriad spiders, and dozens more species of arthropods, mollusks, and annelids that depend on oak leaf litter for nourishment and protection.

Unfortunately, the diverse web of life that is associated with oaks goes unnoticed and thus unappreciated by most homeowners and even many trained biologists. In fact, too many homeowners cut down the oaks on their properties because they have grown tired of raking leaves. The cause of this indifference is lack of knowledge. How can we be interested in or understand the ecological significance of something we know nothing about? Today our ignorance of natural history is a cultural norm. Our attention has been usurped by the digital age, and any spare moments we have each day are consumed by our personal devices or flat-screen TVs. Our school curriculum does not fill this knowledge void, and most assignments have a digital component that teaches us little about the life around us. I meet intelligent adults today, people who have excelled at all levels of their education and are successful members of our society, who cannot even recognize an oak leaf let alone tell me anything about the food webs linked to oaks or the many ways oaks provide the life support we call ecosystem services. Even worse, they fail to see the importance of such minimal knowledge of natural history.

And that is why I've decided to write this book. My point is simply this: there is much going on in your yard that would not be going if you didn't have one or more oak trees gracing your piece of planet earth. But you will never know it unless you have a guide. A similar book could be written about pines, or cherries, or elms, or birches; every tree genus, in fact, has a unique and fascinating story to tell, but those tales will not be as impressive as the story about oaks. Oaks support more forms of life and more fascinating interactions than any other tree genus in North America. All this life does not show up at the same time nor stay with your oak the entire year. Some species, in fact, are so ephemeral you need to be on hand the single day they visit your tree. To fully appreciate what an oak tree can bring to your yard, and into your life if you are willing, we need to follow what is happening on your oak trees month by month through all four seasons.

And that is how I have organized this book: a monthly hint at the many parts of nature associated with the oaks in your yard. I start with October, not because that is the most rewarding month for oak-watching, although it is a good one. Rather, it was October when I decided to write this book.

October

WHEN OUR PROPERTY was no longer mowed for hay, the invasive plants from Asia—multiflora rose, autumn olive, Japanese honeysuckle—that had been kept at bay by the mowing exploded everywhere. These species are incapable of supporting the wildlife we hoped to attract to our new home, so a typical weekend at the Tallamys' involved whacking out the root balls of huge multiflora rose bushes with my mattock. It was satisfying work, especially when we got to stack the whacked bodies up in giant piles, but it did leave open ground wherever we had removed a bush. The following spring we were thrilled but somewhat befuddled to find seedling white oaks and beeches popping up in many of these disturbed areas. Befuddled because I couldn't for the life of me figure out how they had gotten there. We had no white oaks or beeches on our property and no mature trees nearby from which squirrels could have moved seeds even if they had carried them a good distance. I knew that acorns and beech nuts, unlike the seeds of many other plants, do not remain viable in the soil for more than a year, so there was no chance they had sprouted from a seed bank many years old. I was stumped!

An ancient mutualism

I love such ecological puzzles and blame them for all the times I walk down the hall but have no idea why I am walking down the hall. My mind is usually elsewhere. Fortunately, I hadn't puzzled over the oak seedling conundrum too long before I saw a picture in a photography magazine of a blue jay flying with an acorn in its beak. A quick literature search revealed that I had indeed stumbled on the answer: it was blue jays that were bringing acorns and beech nuts to our property and planting them for us everywhere the soil was disturbed enough to make tapping the seeds below ground easy. Although blue jays are the only jays in much of the eastern United States, there are eight species of jays in North America and 40-some worldwide. They all share a common ancestor that evolved about 60 million years ago in what is now southeast Asia, the same time and place that oaks evolved. Oaks and jays are thought to have hit it off right from the start; oaks made large nutritious seeds that are the perfect size and shape for jays to eat, and, in their attempt to store these seeds for long periods, jays became the quintessential acorn dispersers. Over the eons, jays became so dependent on oak acorns that they adapted both physically and behaviorally to acorn traits. The small hook at the pointed end of a jay's beak is designed to rip open an acorn's husk, and a jay's expanded esophagus (its gular pouch) enables it to carry up to five acorns at once while in flight.

Just because a jay can carry more than one acorn at a time does not mean it takes them all to the same place (Bossema 1979). Whereas many birds cache groups of seeds for use during periods of drought or cold, jays bury them as singletons, just beneath the surface of the ground, at sites dispersed throughout their winter territory. These can be over a mile from the acorn-bearing tree, which makes jays the undisputed champions among acorn dispersers and explains where blue jays had found acorns to bring to our property; there are a number of large white oaks and beeches within a mile of our house. The idea, of course, is that each jay will remember all the places it has buried an acorn and then know exactly where to go to retrieve the seed for food as needed during the winter. But apparently this is more of a mental challenge than most jays are up to. And who can blame them;

Blue jays routinely carry acorns for winter storage over a mile from the parent oak.

a single jay can gather and bury up to 4,500 acorns each fall, but it typically remembers where only a quarter of them are buried before springtime. And if a Cooper's hawk manages to eat a jay in December, that jay retrieves none of its acorns. The end result is that each jay plants somewhere in the neighborhood of 3,360 oak trees every year of its seven- to 17-year life span! It's no wonder that jays have enabled oaks to move about the earth faster than any other tree species.

Moving acorns away from the parent tree—where they would surely lose in the competitive battle for light, nutrients, and water—provides oaks with an enormous ecological service, but oaks may derive another important benefit from their relationship with jays beyond the obvious advantages of seed dispersal. Like all organisms, oaks have always had to contend with diseases that attack them over their long life span. In recent decades, however, oaks have been besieged by new diseases introduced from other continents. Sudden oak death and oak wilt are two that have hit oaks hard in several areas of the country. In both cases, however, a small percentage of the oaks in natural and planned landscapes have shown some degree of resistance to these diseases. When disease hits an area, it is the resistant oaks that produce the most and best acorns, and it is jays that preferentially disperse those seeds bearing resistance genes. Although many oaks die quickly from introduced diseases, the seeds of those that remain are spread throughout the countryside by jays, assuring that future generations of oaks will be better able to survive infection. This is natural selection at its best but only works when the partnership between oaks and jays is thriving. The white oak I planted as an acorn in our front yard produced its first acorns last year, so now the ancient coevolved relationship between jays and oaks can play out right in our yard.

Jays, of course, are not the only birds that love acorns. Acorns are a critical component of winter diets for turkeys, Lewis's woodpecker, and many ducks (especially the beautiful wood duck), and they are taken opportunistically by tufted titmice, bobwhite quails, red-bellied wood-peckers, yellow-shafted flickers, eastern towhees, American crows, and white-breasted nuthatches. For the most part, these birds eat acorns as they find them, but a notable exception is the acorn woodpecker of western oak woodlands. Acorn woodpeckers are specialists that store hundreds of acorns for winter use in individual holes they have drilled into snags. They are a colonial species that uses the same acorn storage tree year after year. A sin-gle acorn snag can accumulate 50,000 holes ready to receive acorns each fall. If you live in the West and are lucky enough to have an acorn snag within view of your house, you can cancel your subscription to Netflix; watching your acorn woodpeckers work your tree will be entertainment enough!

The list of mammals that rely on acorns for winter forage is also a long one. We all know how much gray squirrels love acorns, but so do red and flying squirrels, chipmunks, rabbits, black bears, white-tailed deer (acorns make up as much as 75% of a white-tailed deer's diet in late fall), opossums, raccoons, white-footed mice, and voles. And no wonder! Acorns contain large amounts of protein, carbohydrates, and fats, as well as calcium, phos-phorus, potassium, and niacin. If it hadn't been for a healthy supply of

Unlike blue jays in the East, the strikingly beautiful acorn woodpecker stores acorns in pockets it chips out of oak bark throughout its range in western states.

acorns from various oak species, many of the animals just listed would have been devastated when American chestnuts disappeared from eastern forests after the introduction of chestnut blight from Asia.

Masting

Speaking of acorns, have you ever noticed that every once in a while oaks produce an outsized crop of acorns? And it's usually not just one oak here and there, but often nearly all the oaks in an entire region that produce an extraordinary number of acorns in the same year. An exceptional example was seen in the fall of 2019 when members of the red oak group from Georgia through Massachusetts synchronously produced vast numbers of acorns. This phenomenon is called masting, and it has begged explanation for centuries. When I was in graduate school I was taught that masting is an adaptation against acorn predation. Acorns are such a valuable source of food for so many types of animals that if oaks predictably produced a moderate number of acorns each year, the squirrels and the deer, the mice and the jays, the ducks and the towhees, and all the other creatures that rely on acorns to get them through the winter would increase their population sizes to meet the available food supply. This would not be good news for oaks, because every year, large populations of acorn-eaters would end up destroying nearly every available acorn, and oak reproductive success would plummet. But if oaks unpredictably and synchronously produced many more acorns than there were acorn predators to eat them—that is, if they produced a mast—some acorns would escape the predatory scramble for acorns and germinate.

There is a second advantage for oaks that mast, whether it is by adaptive design or simply a fortunate consequence of masting that evolved for other reasons (Koenig and Knops 2005). During mast years, there is unlimited food for acorn predators. This removes one of the biggest factors that limits population growth, and thus birds, squirrels, mice, deer, and so forth typically make more babies successfully during mast years than at other times. Unfortunately for all these hungry new mouths, the year following a mast year is typically (but not always) a bust year for acorn production; the number of acorns drops below that of most non-mast years, and many acorn predators perish. This boom-or-bust approach to acorn production helps keep acorn predator numbers well below what would make oak reproduction iffy during most years.

A second explanation for oak masting has nothing to do with outpacing acorn predators. There is evidence that masting evolved to improve pollination. Oaks within particular taxonomic lineages may synchronize their reproductive effort to maximize pollination efficiency (Pearse et al. 2016). For the most part, oaks are wind-pollinated, and wind-pollinated plants, not surprisingly, are at the mercy of the vagaries of the wind. Simple probability statistics tell us that pollination success will increase if there is more pollen blowing around when the female flowers on oaks are mature and open. Synchronizing the release of pollen in some years, therefore, results in lots of successful pollination and a mast crop of acorns. Why don't oaks synchronize pollen release all the time? Well, don't forget those predators, which would respond by increasing their reproductive rate if masting were predictable. But even if oaks "tried" to synchronize pollen release every

Whether to allocate limited resources, to outpace acorn predators, or to increase pollination efficiency, oaks like the white oak produce a bumper crop of acorns every so often in a process called masting.

year, the time window during which pollen production could be synchronized with other oaks and with the maturation of female flowers is very short. If, by chance, it rains or is unseasonably cold just when female flowers are ready to receive male pollen grains, many flowers would go unfertilized and acorn production would be low (Kelly and Sork 2002).

There is yet a third hypothesis that offers an explanation for masting in oaks, and it has to do with energy allocation (Ostfeld et al. 1996). In most years there are not enough resources (water, nutrients, and sunlight) to produce the energy required for oaks to both grow and make a lot of acorns all at the same time. Making acorns requires a great deal of energy, so during mast years, oaks grow very little. This is evident if you closely examine oak growth rings after the fact. Accordingly, the energy allocation hypothesis suggests that oaks partition available energy; some years they allocate it to growth, other years they direct energy toward reproduction.

It may be obvious by now that, like so many ecological explanations, these three hypotheses are not mutually exclusive. Oaks could simultaneously enjoy selective advantages from masting behavior to allocate limited resources, to outpace acorn predators, and to increase pollination efficiency. You can decide for yourself which explanation makes the most sense as you watch all the wildlife activity around your oak the next time it masts.

November

EACH FALL THE ACORNS dropping from your tree create a "get 'em while you can" feeding frenzy among your local wildlife. If for some reason you ever feel the urge to compete with the squirrels, jays, chipmunks, and other creatures for a few handfuls of acorns, put them in a plastic bag. After a day or two, you will likely notice small, cream-colored, legless insect larvae accumulating in the bottom of the bag. Closer inspection will reveal a small hole in a large percentage of your acorns. The Sherlock Holmes side of your personality will correctly deduce that the larvae have tunneled out of your acorns through those tiny holes. If you then place a larva on some loose dirt, you can watch it wiggle itself beneath the surface and out of sight in less than a minute.

Protein supplements

What you have witnessed is one of the most abundant creatures taking advantage of your oak tree yet one that is most easily missed: the acorn weevil, any one of some 22 species in the genus *Curculio*. Weevils are fascinating insects and belong to the largest family of animals, the Curculionidae, containing at least 83,000 species worldwide. With so many species, you would think we would be bumping into weevils every day. We don't though because most adult weevils are nocturnal, and most weevil larvae are subsurface feeders: they spend their entire larval development inside a seed like an acorn, chestnut, or hickory nut, or tunneling into a root underground. If you *do* encounter an adult weevil, there is no mistaking it for something else: weevils appear to have huge noses! So huge, in fact, that it can be longer than the weevil's entire body. In reality, weevils have no nose at all; what appears to be its nose is actually an elongated extension of the head capsule itself. The weevil's mouth, bearing tiny mandibles, is at the very tip of that extension.

And this may be the adaptation that has allowed weevils to be the most speciose family in the world. The evolutionary innovation of the oddly extended head capsule has provided weevils with a tool no other beetle possesses—a drill, in effect, that gives immature weevils access to food that is beyond reach of predators and parasitoids. Here's how it works in acorn weevils. When a female is ready to lay eggs, she finds a developing acorn, typically in mid-July, and uses her "nose" to chew a tiny tunnel right into the heart of the seed. When the tunnel is complete, she turns around and deposits an egg at the surface of the hole and then plugs the hole with her feces. When the egg hatches, the larva wriggles down to the end of the tunnel, where it eats and grows on the acorn's innards for the next two months. Once the acorn falls from the tree, it's a race against time for the weevil larva; it must get out of the acorn and into the relative safety of an underground chamber before something eats the acorn and ends the weevil's short life. If

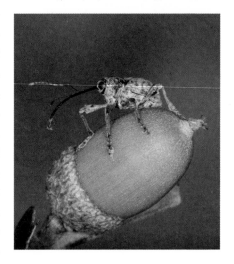

In some years, acorn weevils will lay eggs in 30% of the acorns produced on a tree.

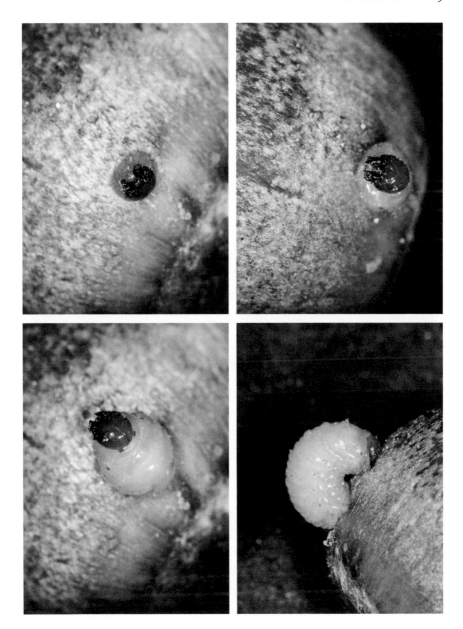

An acorn weevil larva chews a hole in the side of its acorn
and then wriggles through it so it can pupate within the soil.

an acorn bearing a weevil larva *is* eaten by an acorn predator, it's too bad for the weevil, but good news for the consumer, for acorn weevils boost the protein content of acorns considerably.

Even after it chews an exit hole out of the acorn and drops to the ground, the weevil is in danger from shrews, mice, and dozens of arthropods that would consider a chance encounter with this soft ball of protein and fat a lucky event indeed. If the larva succeeds in getting a few inches underground, it stretches up and down, twisting and turning in every direction until it has formed a comfy underground chamber where it will molt into a pupa and remain for the next two years. Somehow the weevil pupa knows when two years have passed; it then molts one final time into its adult form. If someone or something has not stepped on the soil above the chamber, compacting it so tightly that the adult weevil is trapped forever, the weevil will dig its way to the surface, find a mate and a developing acorn, and repeat the process once again.

Acorns, weevils, and ants

Nothing is wasted in nature, including acorns that have already been used by acorn weevils. Acorns are the perfect shape to provide housing for colonies of tiny *Temnothorax* ants and are large enough to hold 100 or so ants comfortably. Gaining access to the cavity within an acorn, however, is a formidable challenge for ants half the size of a grain of rice. Fortunately for these ants, when an acorn weevil exits its acorn, it leaves a hole just big enough for the ants to come and go but not so big that ant predators can easily get into the acorn as well.

It might seem as if *Temnothorax* colonies that have found a cozy acorn with a ready-made predator-proof front door have few ant worries, but not so, for several *Temnothorax* species are what is known as slave-making ants. That is, they raid and enslave other nearby *Temnothorax* colonies. Once through that convenient door too small for predators but not too small for themselves, the slave-makers throttle the queen, kill the workers, and then kidnap the pupae of the victim's colony. When those pupae become adults, they serve at the pleasure of their masters, rearing brood and foraging for

ABOVE

When acorn weevil larvae tunnel out of their acorn, they leave a hole just the right size for *Temnothorax* ants to turn the acorn into a house.

LEFT

Temnothorax longispinosus adults move their larvae into a newly abandoned acorn.

food the rest of their lives. If you have the patience and the eyesight, you can watch such trauma unfold beneath your oak tree from late fall through winter and on into spring.

Blastobasis

Following a very similar sequence of events, the acorn moth (*Blastobasis glanulella*) is another insect that develops within acorns, although it is not nearly as numerous as acorn weevils. I speak of *the* acorn moth, but it is more accurate to think of these tiny moths as a complex of species so similar in appearance that it takes DNA analyses to reliably distinguish among them. Acorn moths do not have a tunnel-digging "nose" to get their larvae into the heart of the acorn, so they use a different, perhaps less novel approach: eggs are laid on the surface of a developing acorn, and the neonate larva (one that has just hatched from its egg) simply eats its way into the nutritious nut. Adult *Blastobasis* moths can be found at lights from April through September, the exact flight period probably reflecting the species and the type of nut—be it acorn, chestnut, or hickory nut—that its caterpillar developed within.

Acorn moths have specialized on using acorns for larval development.

December

IN DECEMBER, my white oak stands out from most of the other deciduous trees on our property in one very striking way: although its leaves are no longer green, they are still on the tree. What's more, they will remain so in diminishing numbers until April! My oak is exhibiting leaf marcescence, the retention of dead leaves on the tree well after other trees are completely bare. Marcescence is common in the Fagaceae, the tree family that includes oaks, beeches, and chestnuts, but it occurs sporadically in a few other genera and even among some tropical trees. Marcescence is particularly noticeable on younger oaks, which often remain fully cloaked in their fall leaves throughout the winter.

Marcescence is odd behavior for temperate zone tree species, and anything that is odd or uncommon in nature is a brainteaser for ecologists. Why do some trees retain their leaves when the vast majority drop them in the fall? It's not surprising that ecologists have proposed a number of possible yet very different explanations, and in this case, all are very difficult to test. It's also not surprising that this level of uncertainty in science frustrates many people. "Just tell us why oaks hold their leaves—no beating around the bush!" For good reasons, we humans like explanations to be clear-cut: black or white, not gray; right or wrong, not "it depends." In the old days, if we saw a saber-toothed cat, those of us who categorized it without ambivalence as "danger" lived to reproduce more often than those who stood around and debated whether or not the cat would attack. But, like it or not, nature is complicated. Even more frustrating is that it is far more common for natural phenomena to have multiple rather than single causes, and many of these selective advantages can be operating at the same time. So, ecologists do the best they can to explain the natural world while they devise bigger and better experiments to test their hypotheses. As Steve Carpenter of the University of Wisconsin once said, "Ecology isn't rocket science; it's much more difficult." So true; so true!

One leading hypothesis about the advantages of marcescence involves browsing mammals, species such as deer, moose, and elk that eat the shoot tips of woody plants. In most places today, those mammals have been reduced to one or two species, but it wasn't that long ago that at least 12 species of large browsers were common in North America and many more throughout the global range of oaks. Claus Svendsen (2001) suggests that when leaves are retained around nutritious buds, it makes it more difficult for browsers to eat buds without getting a mouthful of dead leaves, which are decidedly poor in nutrition and taste. It is simultaneously possible that browsing on branches holding dead leaves simply makes more noise than predator-conscious ungulates are willing to risk (Griffith 2014). Both of these hypotheses are supported by the distribution of marcescent leaves within trees.

Marcescence is typically only a feature of lower branches, ones that would have been within reach of hungry browsers. To this you might say, "My oak is 20 feet tall, and it is covered with marcescent leaves right to the top, way higher than any deer can reach." Quite so, but mastodons were 10 feet tall and could reach another 10 feet with their trunks. Mammoths were even bigger, often reaching 12 feet tall. Giant ground sloths (*Megatherium* spp.) were also 12 feet tall and could reach several feet higher. Only the upper branches of mature trees, the ones with few marcescent leaves, were relatively safe from such huge browsers.

The leaves of young oak trees are marcescent; that is, they remain on the tree in fall and into winter.

Another hypothesis suggests that marcescence is an adaptation that helps trees grow on nutrient-poor soils. Oaks and beeches, our two most common marcescent genera, often outcompete trees from other genera on dry, infertile soils. The idea here is that marcescent leaves trap more snow, increasing soil moisture beneath the tree in the spring when tree growth is fastest (Angst et al. 2017). Moreover, by holding their leaves all winter, marcescent species slow the rate of leaf decomposition, so when the leaves fall in the spring they create a nutrient-rich mulch beneath the tree when the tree needs it most. Remember, though, that none of these hypotheses exclude the others; marcescent trees could easily suffer less from browsing, trap more snow, and grow faster on poor soils all at the same time.

January

ONE COLD JANUARY day when light snow was falling I noticed a small bird flitting about the branches of my oak. A closer look with my binoculars revealed a golden-crowned kinglet meticulously inspecting the smaller branches of the tree. More than once, it picked up something and ate it. I was watching what should have been one of those ecological puzzles I so enjoy, but I hadn't thought about it enough to find it remarkable. After all, my Sibley bird book tells me that at least some golden-crowned kinglets migrate south in the fall from their breeding grounds in Canada. But they don't fly to the tropics like so many other migrants; some, in fact, invariably spend the winter on our property in Pennsylvania. So, here was a bird that is supposed to be in my winter yard; nothing puzzling about that. What should have caught my attention, though, is that golden-crowned kinglets, as well as ruby-crowned kinglets and brown creepers that share similar winter habitats, are entirely insectivorous. That is, rather than eating the seeds that sustain most winter birds, they eat insects and spiders all day, every day. And not just a few insects and spiders, but hundreds each day, particularly in the winter. Because they are so tiny (2 ounces, 3 inches long—only two-thirds

Golden-crowned kinglets eat overwintering caterpillars all winter long.

the size of a chickadee!), they need lots of food to generate enough heat to stay alive in sub-zero nights.

Last winter, I received an email from a recent graduate student who had just dissected an owl pellet (the regurgitated remains of a recent meal), coughed up the night before by the barred owl who showed up most evenings on a fence post in her yard. She sent an image of the pellet's contents, arranged in neat piles by the type of prey the owl had eaten. On one side was a pile of small rodent bones, exactly what you would expect to see in an owl pellet. But on the other side of the image was an even larger pile of caterpillar remains, mostly head capsules that could not be digested. The barred owl had been eating caterpillars in Massachusetts in the middle of January!

Caterpillars in winter

Like my student's barred owl, the tiny kinglet I had been watching must have been eating insects on my oak tree in the middle of January, while the temperature hovered around 20 degrees Fahrenheit. But as any entomologist with over 40 years of immersion in the world of insects (like me) will tell you, there *are* no insects on oak branches in the middle of January. But any entomologist who does tell you that is dead wrong.

The assumption that there are no leaf-eating insects sitting on bare winter branches is based on logic. Not only is it too cold for such insects to function, there is also nothing for them to eat from October to April. Some insects, though, are just not logical! My encounter with the kinglet is one of the things I find so attractive about the study of natural history. "You learn something new every day"—that old cliché must have been coined for nature lovers because it is so true. Its author might well have added, "The more you learn, the more you realize there is to learn." The truth about oak insects in winter became evident a few weeks after I had watched the foraging kinglet when I came across a study by Bernd Heinrich (Heinrich and Bell 1995). Heinrich is one of a dying breed of great naturalists, someone whose observations have so often explained some aspect of the natural world that we mere mortals have never even noticed. I can't remember why, but Bernd decided to dissect golden-crowned kinglets that had been killed during the winter by window strikes in Maine. To his surprise and my amazement, their tiny crops were jam-packed with caterpillars—caterpillars the birds had eaten the same freezing cold day that they had smashed into a window and died!

It turns out that many species of moths, particularly inchworms like the common lytrosis (*Lytrosis unitaria*), spend the winter in the caterpillar stage. They eat green leaves from their host tree in early fall, reaching their third or fourth instar (a little more than half grown) before the leaves turn brown. At that point these caterpillars stop eating and either hide among the nooks and crannies of the tree's bark, or stay where they are on small branches and do absolutely nothing. That's right, they just sit on a twig looking every bit like a stick themselves, all winter long! When it drops below freezing, they rely on glycerin, the same chemical that we use to make car antifreeze, to keep their cells from bursting. The branches of trees in winter that look so bare are actually overwintering sites for caterpillars that sustain insectivorous birds like kinglets and brown creepers, as well as nuthatches, chickadees, titmice, and several woodpecker species that supplement a seed-heavy diet with insects. In fact, 50% of the diet of the chickadees that I had always assumed were entirely granivorous during the winter is insects! This is a cautionary note for us bird lovers: simply keeping our feeders filled with sunflower seeds during the winter is not sufficient to sustain even our most common birds. We also have to plant the tree species that support the caterpillar stages of the moths that seed-eating winter birds depend on all winter long. As we shall see, oaks do this better than any other tree genus in most U.S. counties.

What do birds eat?

Many of us carry misconceptions about nature all our lives. These were created by misinformation that we picked up from our parents, our friends, Walt Disney, Marlin Perkins, or our culture. One such misconception is that if you want to help birds, you give them bird seed. This is actually good advice for a few bird groups that are granivores all year long, such as the finches and doves. They can derive almost all the fats and proteins they need just from seed and have the unusual ability to rear young on regurgitations of a milky substance the adults produce directly from seeds. However, most song birds in North America are primarily insectivores, particularly during that all-important nesting period, with seeds and berries only supplementing their diet. Their nestlings are unable to digest seeds at all; thus, most of our bird species cannot reproduce without a ready supply of insects.

Another misconception is that we don't need to worry about the local supply of insects because insects are everywhere all the time. If this is true, where do all those insects come from and what do they eat? Do they just appear out of nowhere? Is Aristotle's theory of spontaneous generation alive and well in our popular culture? I hope not. The fact is that all insects, every last one of them, are produced directly or indirectly by plants. They either eat some plant part, or they eat another animal that ate some plant part. What follows logically, then, is that when we reduce the amount of plants in any given place, we reduce the diversity and abundance of insects.

Alarming headlines from around the world are reminding us of the critical linkage between plants and insects; we have removed more than half of the forests on earth and, not surprisingly, insect populations have declined globally by at least 45% since 1979 (Dirzo et al. 2014). And again, it should be no surprise that with insect declines come bird declines. There are now 3 billion fewer birds in North America than there were just 50 years ago (Rosenberg et al. 2019), and over 430 bird species in North America are declining so rapidly that they are now considered at risk of extinction (2016 State of the Birds Report). If we think of insects not as creatures with six legs, but as food for birds, amphibians, reptiles, and mammals, we can start to appreciate the ecological significance of insect declines and why we must reverse them.

Like most American birds, this northern cardinal rears its young on insects and spiders.

What makes insects

In contrast, many people *do* know that plants are necessary to produce insects, but they then make the mistake of assuming that all plants produce equal numbers of insects. It's too bad that this isn't even close to being true, for if it were, we could plant eucalyptus for lumber in Portugal, India, and Colombia without fear of creating lifeless rows that we falsely call forests. We could use Mexican pine trees for shade in Peruvian coffee farms without starving the birds that need those farms for wintering habitat. We could move ornamental plants all around the world with no worries of disrupting food webs, and when those same plants escaped our gardens and spread into natural areas as invasive species, there would be no threat of the invaded sites becoming so degraded that they could no longer play vital ecological rolls. Alas, we have not been not so lucky. There are enormous differences among plants in their ability to support insects, and these fall into two categories: differences between native and non-native plants, and differences among native plants themselves.

The differences between native and non-native plants are easy to explain. Because plants lace their leaves with nasty-tasting or toxic chemicals to keep herbivores from eating them, insects that eat plants have had to adapt to these chemical defenses both physiologically and behaviorally. This has been such an evolutionary challenge, however, that most insects have only been able to circumvent the defenses of one or two plant lineages that share common defenses. In other words, most insects have become very good at eating a few plants but are completely unable to eat most plants. We call this "host plant specialization," and it describes how nearly 90% of insect herbivores interact with plants (derived from Forister et al. 2015).

If you are wondering how scientists arrived at this figure, wonder no more. Host records in the literature have allowed us to categorize any given caterpillar species for which we have host data as being a specialist or a generalist. How you categorize your caterpillar will depend on how strict your definition of host plant specialization is, but here are some statistics that may help you decide how to do this. There are 12,810 species of moths and

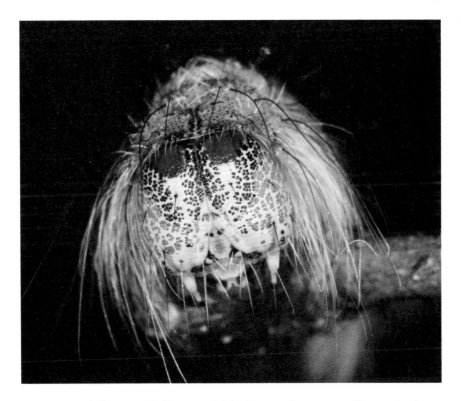

The greater oak dagger moth (*Acronicta lobeliae*) is one of many caterpillar species that can develop only on oak leaves.

butterflies in North America, but only 6,752 have confirmed host records. That's right: we still don't know which plants some 6,058 species of caterpillars eat! Out of those with host records, 86% confine their larval development to plants within three families. This has been the formal definition of host plant specialization in the literature since the '70s. You may think eating plants from three different families is not very specialized, but when you consider there are 268 plant families in North America, being able to eat just 1% of them does indeed sound specialized. Moreover, 67% of our species with known host plants eat plants from only one taxonomic family (or 0.3% of the available plant families), and 49% eat plants in only one genus (or 0.04% of the 2,137 plant genera that occur in North America).

Even our most generalized species, the Io moth, eats only 120 plant genera, a mere 5.6% of the plant genera in North America! The data are clear: nearly all caterpillars are able to eat only a few of the plants species that occur in North America.

Here is one example, although there are many thousands. The greater oak dagger moth (*Acronicta lobeliae*) has, over the eons, become an oak specialist, able to eat tough oak leaves without ill effects from the phenolics, tannins, and lignins that characterize oak defenses. But in specializing on oaks, this dagger moth did not spend any evolutionary time adapting to the cardiac glycosides of milkweeds, the salicylic acid of willows, the aristolochic acid in pipevine, the juglone in black walnuts, and so on. It can develop only on oaks. So, if I replace my oak with a Callery pear, golden-raintree, dawn redwood, or any other tree from Asia or Europe that produces phytochemicals that are different from those of oaks, the greater oak dagger moth would not even recognize it as a plant and would die if it were somehow forced to eat it. Without my oak tree, this cute shaggy caterpillar would disappear.

Keystone plants

It is a bit harder to explain why there are huge differences in insect productivity among native plants. I am not exaggerating when I say huge: the number of caterpillars hosted by North American native plants varies in any one location from well over 500 caterpillar species (*Quercus*; oaks) to no caterpillars at all (*Cladrastis*; yellowwoods). We can measure the ability of plants to host insects by simply counting the number of species of caterpillars that scientists have found eating particular plant genera over the years. Caterpillars are used for these measures because host plant data are much more complete for moths and butterflies than for other groups of herbivorous insects. What's more, the number of caterpillars supported by plant genera do not form a smooth continuum from few to many; instead, in any one place, just a few plant genera, about 7%, support most of the caterpillars, and most plant genera support only a few caterpillar species. To put it another way, about 75% of the insect food required by birds and other animals is produced by just a few plant genera. In most U.S. counties, oaks, cherries, willows, birches, hickories, pines, and maples are producing the vast numbers and types of insects that support animal populations. These tree genera are keystone plants because they play the same support role that the keystone in a Roman arch plays. In place, the keystone supports all the other stones in the arch, but take the keystone away and the arch collapses. The same is true for keystone plants; if you have an oak or black cherry or black willow in your yard, there will be enough insect food for at least a few birds to reproduce. But a yard without keystone plants will fall far short of the insect abundance necessary to sustain viable food webs, even if dozens of native plant genera are present.

And this is why the oak in my front yard is not just another tree. In my county in Pennsylvania, 511 species of moths and butterflies develop on oaks—nearly 100 more species than their closest competitors, the native cherries. No other tree genus supports so much life. There is nothing unusual about my county, either; oaks are top life-support trees in 84% of the counties in North America, which is just about every county in which they occur. How do my oaks compare to other native trees in my yard? Really, really well. My maples are powerful trees, supporting a potential 295 species of caterpillars, but not nearly as powerful as my oaks. My white pines are also productive, supporting up to 179 species, but that is only a third of what my oaks support.

Many of my other trees are mere shadows of my oaks in their ability to support the food web in my yard. My ironwood (*Carpinus caroliniana*) can support just 77 species of caterpillars; sweetgum (*Liquidambar styraciflua*), a paltry 35 caterpillar species; and so on. Unfortunately, this is especially true for many natives that are popular as ornamentals. My dogwood (*Cornus florida*) supports 126 species of caterpillars, my serviceberry (*Amelanchier canadensis*) supports 114 species, my redbud (*Cercis canadensis*) supports only 24 species, and my spicebush (*Lindera benzoin*), just 11 species. That's not to say we shouldn't plant these species; there are specialist caterpillars that exclusively use each of these natives, and they wouldn't be in our yards without their host plant. But it is to say that we should not eschew the mighty oak. A yard without oaks is a yard meeting only a fraction of its life-support potential.

Comparing oaks to the introduced plants that form the backbone of the horticultural trade is even more dramatic. Only three species of caterpillars have been found on crepe myrtles, none at all on camellias and zelkovas, and I have found only one species, uncommonly, on Callery pear!

I will close this section with actual species counts that I have made in my yard over the past three years. So far, I have recorded 923 species of moths (I haven't gotten to the butterflies and skippers yet). Of those 923 species, 811 have known host plants; we still have much to learn about what many species eat. Of the 811 species for which we do have host lists, 245 species include oaks in their larval host choices, and 27 species can develop only on oaks. These numbers are probably low because some of the 112 species for which there are no host records will likely turn out to be oak-feeders. We have 59 genera of woody plants on our property, only one of which is *Quercus*, the oaks. My point is, oaks represent less than 2% of our woody plant diversity but support at least 30% of our moth species. In comparison, of the 811 moths I've found so far and whose preferences are known, 129 species (16%) use hickories, 70 species (9%) use viburnums, 49 species (6%) use serviceberries, and just 17 species (2%) use tulip trees. This is good evidence that oaks shine brighter than other plants in their contributions to biodiversity on the national level, the regional level, and even the level of a single yard.

Oaks are best—but why?

The data are clear: oaks support more caterpillars than other plant genera, not just where I live, not just east of the Mississippi, but in most of the United States. But why? What makes oaks so much better at producing the caterpillars that fuel most terrestrial food webs? It's a good question that no one has definitively answered, but a handful of hypotheses suggest possible explanations (Janzen 1968, 1973; Southwood and Kennedy 1983; Condon et al. 2008; Grandez-Rios et al. 2015). When we say more caterpillars use oaks than other plants, we mean that more caterpillar species have adapted to the many phenolics that characterize oak chemical defenses. So, any feature of oaks that encourages such adaptations might help explain why so many species can now use oaks as host plants.

One such feature may be the size of the genus *Quercus*. There are about 600 oak species worldwide, with 90-some occurring in North America, making *Quercus* the largest tree genus within the northern hemisphere. By comparison, there are 400 species of *Prunus* (cherries) and 350 species of *Salix* (willows) worldwide; large genera to be sure, but they still contain hundreds fewer species than *Quercus*. And tree genera like *Acer* (maples; 160 species), *Pinus* (pines; 111 species), and *Betula* (birches; 30 to 60 species) are much more representative of average genus size among most trees.

Highly correlated with genus size is the geographic area covered by a genus. Tree genera that occur over large areas of the country, or, indeed, the world, overlap the ranges of many more caterpillar species than tree genera that are not widespread. This puts hundreds of caterpillar species in contact with widespread tree species for long periods of time, a prerequisite for the evolutionary interactions that lead to host use by caterpillars. And *Quercus* has the greatest geographic range of any tree genus: oaks occur across Asia, Europe, and North America, and even extend through Central America and into the northern portions of South America.

Along similar lines, plant apparency has also been considered a factor in the evolution of host use. Plants that are large, and/or present in the landscape for long periods of time, are more easily encountered by moths and butterflies (caterpillar adults) and are therefore more likely to develop close relationships with Lepidoptera species than tiny plants with very short life cycles. An oak, for example, exists in exactly the same spot year-round for hundreds of years and becomes an enormous individual over that time span—one that is easy to find if you are a female moth loaded with eggs. It is far more likely that local lepidopterans would adapt to oak tree defenses than to a tiny and ephemeral spring beauty (*Claytonia virginica*), whose total biomass is less than that of a single oak leaf and whose entire life cycle is completed in a few weeks.

Yet another feature of the genus *Quercus* is its age. Oaks evolved in southeast Asia about 60 million years ago. They have been in the New World for at least 30 million years, which has provided New World insect lineages lots of opportunity to adapt to oak defenses. Yet, for me, this is not a very convincing explanation for the large number of caterpillar species associated with oaks. There are many other tree genera with ancient origins that do not support many caterpillar species at all. *Liriodendron* is a classic example; older than the genus *Quercus* by millions of years, the tulip tree (*L. tulipifera*) is the only extant species of this genus in North America. Despite evolving sometime before the late Cretaceous, the tulip tree supports only 29 species of caterpillars compared to the hundreds supported by oaks.

Finally, the type of chemical defenses used by plant lineages has been implicated in influencing the rate at which insects adapt to plants. Defensive chemicals have been classified into two groups: quantitative defenses and qualitative defenses. Quantitative defenses are chemicals that are not immediately toxic to consumers but become more effective through repeated exposure; that is, they work best as they accumulate in a caterpillar's body. The tannins produced by oaks are an excellent example of a quantitative defense. They do not poison a caterpillar when eaten but instead impede protein assimilation. This is a good defense because plant leaves contain very little protein even under the best of circumstances, and leaf-eaters cannot afford to lose any of it because of tannins. Unless caterpillars have developed physiological adaptations to counter the effects of tannins, the more oak leaves a caterpillar eats, the less protein it assimilates from those leaves.

In contrast to quantitative defenses, qualitative defensive chemicals are immediately toxic and typically require that caterpillars evolve specialized physiological adaptations, such as the acquisition of particular detoxifying enzymes, before they can eat these compounds without dying. Monarchs, queens, and other members of the *Danaus* butterfly lineage, for example, can develop on *Asclepias* milkweeds because they long ago evolved the ability to detoxify, store, and excrete cardiac glycosides, the poisonous compounds in milkweeds. The point is, it is apparently much easier for insects to adapt to quantitative defenses like the tannins in oak leaves than to qualitative defenses like the cardiac glycosides in milkweeds, or the cyanide in cherries, or the nicotine in tobacco, etc. But again, it is likely that all characteristics of oaks—the large size of their genus and its geographic range, their outsized apparency within ecosystems, and their reliance on more easily circumvented tannins as their primary defense—have contributed to the large number of caterpillar species that rely on oaks for growth and reproduction.

February

FEBRUARY MAY BE the quietest time of year for the oaks in our yard. Their acorns have dropped, most of the caterpillars hiding on oak bark and branches have already been eaten, many of the mammals that visit our oaks in good weather are asleep in hibernation, and it is often the time of year that snow lies deepest beneath our trees. Inside our house though, spring fever runs high. Each day the sun rises earlier in the morning and sets later in the day. Our chickadees, titmice, and Carolina wrens start singing their spring songs on warmer days, and seed catalogues appear in our mailbox. All this invariably triggers great planning sessions, where Cindy and I discuss new goals for our property—goals we never quite meet, but they're fun to talk about anyway. I'm assuming other people use February for yard plans too, which makes this a good time to address common misgivings about using oaks as landscape trees.

Oak misconceptions

What gives many people pause about planting oaks is the immense size of many oak species when fully mature. I can't tell you how many times people have told me they can't plant an oak because their yard simply isn't big enough. I have also heard that homeowners don't want to plant oaks because their roots lift up sidewalks and driveways, their acorns trip people, their leaves have to be raked, they're too expensive, and they or one of their limbs will eventually fall on the house or car. Oh, sigh. If that isn't a "glass half empty" perspective on the use of oaks, I don't know what is. It's true that some of these things can and occasionally do happen, but it's also true that most are easily avoided with a little planning.

Let's start with the size issue. If all oaks quickly reached 100 feet tall with a crown spread of 120 feet and a trunk circumference of 15 feet, this complaint would be valid. But this is not the case; most oaks are smaller than that even when fully mature and typically take centuries to reach their full size. Several species are so small, in fact, that they are more like understory trees than large canopy trees; this makes them perfect for the tiny yard. Species like dwarf chinkapin oak (*Quercus prinoides*), Mexican blue oak (*Q. oblongifolia*), and Gambel's oak (*Q. gambelii*) are particularly suited for lots in urban settings, not just because of their intermediate to small size (e.g., dwarf chinkapin oak rarely exceeds 10 feet tall), but also because they do well on dry, infertile soils. Dwarf chinkapin and Gambel's oak even tolerate the highly alkaline conditions so common in cities. For obvious reasons, it's important to choose a species that is a native resident of your area. Gambel's oak, for example, is found throughout the middle and southern Rockies, Mexican blue oak thrives in the Southwest, and dwarf chinkapin oak has a broad distribution east of the Plains states. With over 90 species of oaks in North America, there are oaks that are appropriate for all but the driest and most northern areas of the United States.

Let's say you have a yard that can handle a larger oak species. Will the roots of that oak lift up your driveway or sidewalk? Again, this can happen if you choose a species with a shallow root system, like willow oak (*Quercus phellos*), but many oaks send their roots deep enough that they do not lift up hardscape. Of course, if your yard sits on shallow bedrock, even a dandelion will lift up your driveway (kidding), but in typical yards, this is not a concern with deep-rooted oaks like northern white oak (*Q. alba*), northern red oak (*Q. rubra*), Shumard's oak (*Q. shumardii*), or blue oak (*Q. douglasii*).

Deep-rooted oaks like this northern red oak (*Quercus rubra*) will not lift up your sidewalk, driveway, or roadway, even if planted right next to them.

Are oaks too expensive? When you make the classic mistake of buying a large tree, they surely are. But an acorn is free, and a seedling is only a few dollars. I am not recommending that you forgo instant gratification and start small just to save money, although for many of us, that's a darn good reason. I *am* recommending that you start with the youngest oak you can find because you will end up with a far healthier tree that will catch up with and even outstrip, in size, any larger specimen in short order. When you buy a large tree, and by large, I mean 1-inch caliper or more, the tree cannot be moved without being severely root-pruned. Or, if it has been grown in a pot, its roots are wound around each other so tightly (rootbound) that they will strangle each other as they grow. Those precious roots—which allow oaks to grow well each year, outcompete other young trees, withstand serious drought, and fight disease—are chopped literally to within an inch of the tree's life, or bunched into an unnatural tangle that will considerably shorten the life of the tree. Large transplants have a 50% chance of dying within the first few years, and if they do survive, they spend over a decade allocating most of their growth resources toward rebuilding their lost root system. In contrast, young oak trees lay down a large root system, among the largest in the plant kingdom, which will support rapid growth and health for the rest of the oak's life. In just a few years, a young tree with an undisturbed root system will catch up with a much larger transplant or potted tree and exceed it in size. What's more, it will enable you to experience, enjoy, and appreciate its transformation from tiny seedling to dominant tree in your yard.

But will your oak eventually fall and cause property damage—or worse? With increasing news coverage of trees toppling during storms, and more frequent and stronger storms each year, it would be easy to conclude that disaster is lurking behind every tree in your yard. The fact is, large trees do cause disasters on occasion, but there is much we can do to minimize or even eliminate the risks that residential trees pose. One obvious solution is to avoid planting trees that will eventually grow large near the house. As logical as this is, the result would be a treeless zone 100 feet wide around every building, a practice with serious downsides. It would eliminate the cooling effects trees provide in the summer and the warming effects they

This seedling white oak will spend its first year building the powerful root system that will see it through hundreds of years of good health.

can provide in the winter. And much of that treeless zone would probably end up as lawn, a landscape use that destroys watersheds, does not support pollinators or food webs, and, in short, offers little to no ecological value; grass is our poorest plant choice for storing carbon. Fortunately, there is an alternative: plant tree groves rather than specimen trees.

Because trees grown in isolation without competition from other trees for light, water, and nutrients usually grow more massive, achieve a larger crown, and present a grander aesthetic than trees that grow close to other trees, we almost always plant trees as isolated specimens. But this practice invites the very disasters we pray will never happen. In most treed areas of the country, trees grow in forests, not by their lonesome. Yes, each member of the forest is a bit smaller than it would be on its own, but it is also far more stable. Trees that grow at a spacing found in most forests interlock their roots, forming a continuous matrix of large and small roots that is extraordinarily difficult to uproot. When the big winds come, grouped trees may lose a branch or two, or in the extreme winds of a hurricane or tornado,

Trees like these western hemlocks in Oregon typically grow in clusters under natural conditions, which allows them to interlock their roots in ways that protect against blowdown.

their trunks may even snap off some feet up from the base, but they rarely blow over entirely. With this in mind, an easy way to reduce the risks from treefalls in your yard is to plant trees in twos or threes, maybe on a 6-foot center, creating small groves that the eye will take in as if it were a single tree. Trees planted in this fashion weave that stabilizing web of roots that will hold them upright even in extreme weather. There is a catch, however. You have to plant such trees when they are young. In fact, the smaller the better. This way your trees will have every opportunity to interlock their roots as they grow. So now we have two excellent reasons to plant your oaks (and other tree species as well) as small as you can: they will be healthier trees that grow much faster than large transplants, and they will have the opportunity to interlock their roots with close neighbors, reducing future risks to life and property.

March

BY THE TIME MARCH rolls around, most of the marcescent leaves on young oaks and immature oak branches have fallen from the tree, and the greater number of leaves that dropped from mature branches in the fall have melded with those deposited under your oak in previous years to form the priceless organic layer we so inappropriately call litter. It is not at all obvious, but there is far more life under our oak trees than on them. Despite the many hundreds of species of moths and butterflies, katydids, walking sticks, tree crickets, lace bugs, cicadas, planthoppers, treehoppers, and gall wasps that depend on oaks, as well as all the mammals and birds that use oaks in one way or another during their life history, the diversity and abundance of the little creatures that reside in the leaf litter that accumulates beneath an oak tree is astounding and easily exceeds counts in the millions. Many young fishing enthusiasts have learned that earthworms are plentiful beneath oak leaf litter, but in their enthusiasm to grab the wrigglers, they often miss the collembolan springtails, conehead proturans, two-pronged bristletails and other pale diplurans, machilid jumping bristletails, carabid ground beetles, dozens of species of moth caterpillars that eat dead leaves instead of green ones, and the vast assemblage of mites, snails, slugs, centipedes, millipedes, pill bugs, roundworm nematodes, sow bugs, and spiders that together form a complex community of decomposers and their predators.

Many of these creatures are the most abundant multicellular organisms on earth. Among litter arthropods, for example, springtails can easily exceed 100,000 individuals per square meter of litter (Ponge 1997), and proturans run a close second at 90,000 per square meter (Krauß and Funke 1999). In sheer numbers, though, litter mites take the prize for arthropods: in a single square meter of temperate forest leaf litter, there can be 250,000 mites from 100 different families. As abundant as these litter arthropods are, they pale before the size of nematode populations: one square meter of litter and soil humus can contain over 1 million nematodes (Platt 1994), making them the most numerous animals on earth.

Lest mushroom hunters feel slighted by my focusing on the animals that depend on oak leaf litter, it's no secret among fungi foragers that rich oak leaf mold is an ideal place to look for dozens of species, including *Lactarius* milkcaps, brightly colored russulas and boletes, and, of course, truffles and morels, both of which use energy from the roots of oaks as well as their decaying litter.

Priceless litter

To be fair to worm catchers, many of the inhabitants of oak leaf litter are barely visible to the naked eye even though they are everywhere you look. But there is an easy way to observe some of the larger arthropods in the litter beneath your feet: scrape away the top layer of leaves and nestle a white sheet of paper into the spot. After a minute or two your paper will be dotted with dozens of springtails, mites smaller than a period, and whatever other species happen to find the paper in their path. What you are seeing is the vibrant community of decomposers or, in the vernacular of soil ecologists, detritivores—those creatures that derive their nutrition from dead plant parts, or from the bacteria and fungi that help break down the plant cellulose in those parts that is so difficult for animals to digest. And where there are thousands of detritivores, there are hundreds of detritivore

Oak leaf litter provides housing, nourishment, and moisture for more multicellular species than are found aboveground in oak trees.

predators keeping the community in trophic balance. The importance of the contribution these unobtrusive animals make to the greater web of life is unappreciated by most but hard to overstate.

Plants are the only organisms that can capture energy from the sun and convert it to food that sustains the rest of life. All plants require basic nutrients, such as nitrogen, phosphorus, and iron, which they extract from the soil with their roots and use for growth and reproduction. These nutrients, particularly nitrogen and phosphorus, are often present in soil in limited amounts and can easily become depleted if they are not returned to the soil after they are used. As a plant grows, nutrients become locked in its tissues, and unless they are released when the plant or its tissues die, they are no longer available for new plant growth. And that is the essential role of decomposers, to recycle the nutrients needed directly by plants and indirectly by nearly all animals. What is special about the 700,000 leaves that fall each year from a mature oak is that they create the very best type of leaf litter for sustaining decomposer populations.

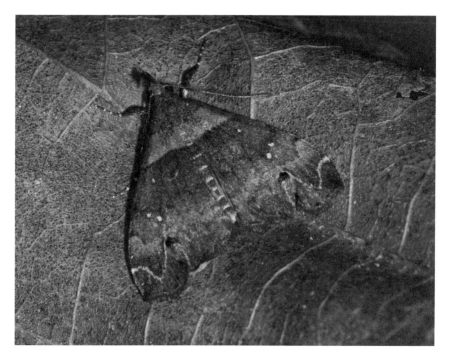

The ambiguous litter moth (*Lascoria ambigualis*) is one of 70 moth species that eat dead rather than live leaves.

There are several reasons oak litter supports decomposers better than most other types of leaf litter. Top among these is its persistence; most species of oak leaves decay very slowly once on the ground, consistently providing the housing, food, and humid conditions required by decomposers for up to three years. Because the litter is replenished annually, there are always plenty of leaves on the ground in various stages of decay under oaks. The same cannot be said for leaves from most other deciduous trees. Maples, tulip trees, birches, aspens, cottonwoods, sweetgums, and hickories, just to name a few, all have thin leaves that decay relatively quickly once off the tree. In fact, the leaves of many trees decay so fast that they do not last through the summer until the next leaf fall. This is a serious problem for decomposers, very few of whom can survive for long on or in soil that is not covered by leaf litter. Not only does bare soil lack the organic material that sustains decomposers, it also rapidly loses the moisture on which the decomposer community depends. If leaf litter disappears, so do the decomposers, as well as the fungi and bacteria many eat, and the mycorrhizae that enable plant roots to absorb the nutrients they need. Oak leaves, in contrast, are loaded with lignins and tannins, which retard breakdown; any leaf pack with a substantial proportion of oak leaves protects the ground and its diverse inhabitants year-round.

Oak leaf litter is proving to have another very practical benefit in this age of invasive species. One of the recent scourges of eastern deciduous forests is Japanese stiltgrass (*Microstegium vimineum*), an invasive plant that grows well in sun and shade and now blankets, and I truly mean blankets, forest floors in New Jersey, New York, Pennsylvania, Connecticut, Virginia, Maryland, and who knows where else. It loves disturbance but does poorly in areas with heavy oak leaf litter. The only places on our property that are stiltgrass-free are areas under our oaks. And it's not only stiltgrass that oak litter repels; three species of Asian worms have become serious problems in soils from the Great Lakes to the Atlantic Ocean. Particularly notorious are jumping worms (*Amynthas* spp.), which wriggle so fast out of the soil they appear to jump. These smallish, reddish worms are terribly destructive to soil ecosystems. They quickly strip the litter layer, leaving bare, easily eroded soil from which nutrients rapidly leach. In fact, they eat all the organic matter in soil, changing the soil pH and consuming even the tiniest seeds. Explosive populations of jumping worms are reducing both plant and animal biodiversity wherever they occur—except in oak forests. Apparently oak litter can be too tough for Asian worms, and they rarely penetrate areas with abundant oak leaves on the ground. In many places, oak-dominated forests are the last refuge for many spring ephemerals like trilliums and trout lilies.

Besides fighting invasives and enabling the decomposers that recycle vital plant nutrients, the persistence of oak leaf litter provides another enormous ecological benefit: it improves water infiltration. The thick mat of leaf litter that characterizes forests with numerous oaks acts like a sponge when it rains and is most valuable when it rains hard (Sweeney and Blaine

2016). The water from a 2-inch downpour, for example—more than 54,000 gallons per acre—is captured almost entirely by an oak forest's leaf litter and the organic humus it creates. Litter and humus don't hold this water indefinitely, but they do corral it on-site just long enough for it to seep into the ground, replenishing the water table on which so many of us depend. In areas with no leaf litter, the same 2-inch rainstorm causes a flood. Bare soil cannot and does not hold water in one spot long enough for infiltration to occur. Instead, rain water rushes off-site, usually taking soil with it during each rain event, causing the soil erosion that clogs our streams and rivers, silts up our dams, and worst of all, eliminates the organic-rich, nutrient-laden topsoil that has stored tons and tons of carbon deposited over the years by plants and their mycorrhizae. This is nothing less than ecological devastation for the soil community, and it takes decades to repair such soils.

There is yet another benefit derived from oak leaf litter. As water that is held on-site by leaf litter seeps through soil pores on its way to the water table, it is purified. Excessive nitrogen and phosphorus loads that come from lawn and farm fertilizers are filtered out of the water, as are heavy metals, pesticides, oil, and other pollutants. The upper layers of the water table are always on the move, slowly sliding underground toward the nearest stream or river. These purified waters enter waterways slowly and evenly, delivering that 2-inch rain in a trickle over days and weeks instead of a single gushing flood of stormwater runoff. This stabilizes water flow in streams which, in turn, prevents scouring and the destruction of aquatic communities of insects, crustaceans, and fish. Who needs healthy stream communities? Well, who needs plentiful, reliable, clean, fresh water? Everything and everybody! Streams with diverse communities of aquatic insects and crustaceans carry two to eight times less waterborne nitrogen than streams with no aquatic arthropods; they also contain higher levels of dissolved oxygen (Sweeney and Newbold 2014). And just as leaf litter sustains the arthropods in terrestrial decomposer communities, leaves that fall into waterways sustain stream arthropods. Grazers and leaf shredders like crayfish, stoneflies, mayflies, and caddisflies rely on the energy locked up in the leaves themselves as well as the algae and diatoms that colonize the surfaces of leaves immersed in streamwater. Oak leaves last as a reliable source of food in streams far longer than leaves from other types of trees, comprising the bulk of the leaf pack in vibrant streams.

Oak leaf memories

Through age six I lived on an oak-lined street in Plainfield, New Jersey. Each fall my father would rake the leaves generated by these oaks from our small yard to the curb—and burn them. Back then, burning seemed like an acceptable solution to the fall leaf "problem." I, for one, loved this annual tradition. It was fun jumping into the huge leaf piles my father had raked up (at least they seemed huge to me) and even more fun poking the leaf piles with sticks once they were alight. Most of all, I loved the smell created by burning oak leaves, and to this day the smell of burning leaves conjures some of my best nostalgic memories. But leaf burning, as well as most other typical landscaping practices, slowly transformed our yard into an ecological dead zone. Gone were the millions of decomposers that our oak leaves could have nurtured, and the nutrients they would have returned to the soil were instead washed down the sewer with leaf ash during the first rain. Neither my father nor his neighbors thought about the simple yet intricate ways oak leaf litter could have contributed to watershed management in Plainfield—how the leaves we could have placed in beds about our yard would have helped to purify our groundwater, prevent destructive floods somewhere nearby, and create a healthy ecosystem in Meadow Brook, the small stream at the end of our street. The concept of ecosystem services had only recently occurred to the most forward-thinking ecologists, and it wasn't close to being part of the public discourse.

It has been 62 years since my father burned our oak leaves and, in that time, we have learned a great deal about what keeps the natural world afloat. Leaf burning is now banned in most towns because of the ensuing air pollution, and more and more people are discovering that leaves, particularly oak leaves, are superior mulch for flowerbeds and trees. Many townships gather leaves that homeowners don't want, compost them, and then offer them for free to homeowners who appreciate their ecological value. And fortunately, each year there are fewer homeowners who cut down magnificent oaks simply because they have grown tired of raking leaves. It has become clear to me that developing a sustainable relationship with the oaks in our landscapes will lead us to a sustainable relationship with the natural world on which we all we depend. Every day I see signs that our society is, in fact, learning how to do this. Yay!

April

FROM THE PERSPECTIVE of oak phenology, April (from the Latin *aperire*, "to open") is aptly named. April is the month that oak buds swell and open, releasing their first flush of young leaves. But it is also when a biological event occurs that is so ephemeral, almost no one has witnessed it. It's not that this event is rare; it typically happens dozens to hundreds of times on each oak of any size and usually in several different ways. Moreover, the location of the event is entirely predictable: it will happen at a leaf bud. But all expanding buds are potential sites for this event, and a good-sized oak can have hundreds of thousands of buds, so choosing the right bud to watch involves more than a little luck. The biggest challenge to observing this event, though, is its brevity. Its exact timing depends on both the local weather and the microclimate where an individual oak is growing. Each bud is vulnerable only for a few hours as it reaches and then passes through a particular stage in its growth, the point when the meristematic tissues within the bud are just at the right stage. I am talking about the precise moment each year that female cynipid gall wasps inject both an egg and hormone-mimics into the undifferentiated tissues of oak buds to form a gall. From start to finish, oviposition within a bud takes about five minutes.

A lot of galls

Cynipid gall wasps do not fly around stinging people, but they are members of the order Hymenoptera: the wasps, bees, and ants. Like the majority of hymenopterans, cynipids are tiny and cannot sting you, and most people would dismiss them incorrectly as midges or gnats—that is, if they noticed them at all. Cynipids are *gall* wasps because they chemically manipulate plant tissues to form galls, plant growths that provide both protection and food for cynipid larvae. There are nearly 800 species of gall wasps in North America, and most of these have specialized relationships only with oaks.

Whereas your timing has to be just right to see a gall wasp initiate a gall, once formed, the gall itself is a conspicuous feature on oak leaves for the rest of the year or, if formed on twigs, for many years. After an egg is laid and bathed in plant growth regulators deposited by mother cynipid, the plant responds with an explosive, cancer-like growth of the cells surrounding the egg and the resulting cynipid larva. The shape of the gall varies with the species of gall wasp that stimulated its formation and can be further manipulated by the gall wasp larva itself, but most galls share common features. The outermost layer is typically quite hard and difficult for casual chewers or gall wasp parasitoids to penetrate. It is also liberally anointed with foul-tasting tannins to further discourage predators. Another common feature is an extremely hard sphere surrounding the inner larval chamber, often but not always at the center of the gall. This layer of gall tissue is a second line of defense that protects the larva within, in case a parasitoid manages to breach the outer wall. Immediately surrounding the larva is a nutritious layer of plant cells that will supply all the food the larva requires until it completes its development. When the larva reaches its full size, it pupates within the same tiny chamber and emerges from the gall as an adult either the following April, or, if it is a species with two generations per year, in late June just before the oak experiences its second flush of leaves.

The sequence I just described represents a highly specialized relationship between cynipids and oaks, but it tells only half the story. Most cynipid species, particularly those associated with oaks, have a complicated life history known as alternation of generations. The first generation is comprised entirely of parthenogenetic females—that is, females that can lay fertile eggs without mating with a male. That's a handy trait to have because there are no males in this first generation. The adults and galls produced by the first generation have a morphology specific to each species. The second generation, in contrast, produces adults and galls that are entirely different from those of the first generation, and instead of just females, it yields both males and females that need to mate in the usual way to produce viable eggs. For the longest time, cynipid taxonomists thought the two generations were two different species, and you can hardly blame them; the cynipids in each generation looked entirely different from each other, as did

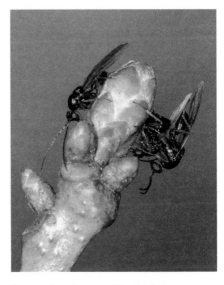

Two male gall wasps (Cynipidae) guard a female as she initiates a gall in a developing white oak bud.

their galls. I'm still not sure how, without the aid of the DNA analyses, taxonomists ever figured out that the wasps that looked one way in April and another way in June—and which produced vastly different galls—both belonged to the same species.

The diversity in gall size and shape is astonishing. I suppose it shouldn't be, though; gall morphology is unique to each species, and most of the nearly 800 North American cynipids make two kinds of galls. That's a lot of gall variation! There are galls that look like small mushrooms, galls that look like sea urchins, and galls that drop off the leaf that made them and jump around on the ground until they wedge themselves into a protective

Oak galls vary with the species of gall wasp that made them and the generation of that wasp.

crevice. There are galls that are the spitting image of Hershey's kisses, polished vases with stoppers in them, medieval maces, apples, pumpkins, pine cones, bullets, and curled-up hedgehogs. In each case, however, there are some adaptive features that help shape the gall, and many of these have to do with escaping the most deadly enemies of gall wasps, cynipid parasitoids (Bailey et al. 2009).

Despite their protective starter homes, gall wasps are among the most heavily parasitized group of animals on earth. Each gall wasp species is successfully attacked by up to 20 species of arthropod enemies, including inquilines that do not make galls themselves but that take over galls made by other cynipids. Mortality from gall wasp parasitoids, in particular, can be extremely high. I call them parasitoids and not parasites because parasitoids are more like specialized predators: they kill their hosts rather than just annoy them. But unlike predators, a parasitoid eats only one prey item its entire life. Just like cynipids themselves, cynipid parasitoids are tiny wasps, but rather than laying eggs in plant buds, they insert their eggs into gall wasp larvae. The egg hatches and the parasitoid larva slowly eats the cynipid larva, starting with the non-vital tissues and then finally consuming the entire creature. In this way, their prey remains alive and fresh as the parasitoid develops. Adult parasitoid females insert their eggs into their prey with an ovipositor, an organ structured like a hypodermic needle. The "stingers" we have grown to fear in much larger social wasps are actually modified ovipositors. In some species of parasitoids, the ovipositor is short; in others, it is longer than the female's entire body. Not surprisingly, its length has been determined by long periods of cat and mouse evolution between the parasitoid and the gall wasp species it attacks.

This can be best demonstrated by taking a look inside a gall. Choose a big gall species like an apple oak gall because it will be easier to manipulate. If you cut the gall in half, the first thing you will notice is that, other than a few thin rays of plant tissue, the gall is largely hollow. You might incorrectly assume that the gall wasp has eaten out the insides of the gall and has already left as an adult, but closer inspection will reveal a circular capsule at the very center of the gall. Cut the capsule open and you will find the tiny cynipid larva occupying only 2 to 3% of the gall's total volume.

What is all that empty space about? It is a simple yet evolutionarily elegant way to keep the cynipid larva just out of reach of its parasitoid species with the longest ovipositor. Over time, strong selection pressure from parasitoids favored cynipids that could make galls large enough, with just enough hollow space between the outside of the gall and the centrally positioned gall wasp larva to avoid their enemies' ovipositors. This, in turn, selected for parasitoids with ovipositors just long enough to reach the center of the gall and thus the cynipid larva within. Over eons, these back and forth evolutionary pressures between cynipids and their many enemies helped create the variety of gall shapes and sizes we see today.

But the large internal space in some galls is just one way that selection has reduced mortality from parasitoids. Some galls avoid parasitoids by growing long, dense hairs on their surface, making it difficult for parasitoids to land on the gall and insert their ovipositors. Others secrete a sticky substance that traps parasitoids if they walk on the gall. Another common strategy is to play hide and seek with parasitoids; galls with urchin-like spikes do not have one central larval chamber like other galls but rather contain a single larval chamber beneath only one of the many spikes. It's then up to the parasitoid to discover which is the spike above the actual galler. Warty galls contain a labyrinth of chambers, but again, only one holds a cynipid larva. Perhaps the most innovative solution to the parasitoid problem, though, is to hire bodyguards. And among insects, the best bodyguards are ants. Ants are omnivores; they readily catch and eat other insects, but they also have a sweet tooth (or mandible). Some gallers take advantage of both these traits by creating galls that secrete a honey-like substance which attracts ants. In return, the ants hanging around the sweetened gall protect it from any incoming parasitoids.

The ability to induce oak tissues to change their normal growth patterns seems like a one-sided relationship that provides housing, protection, and food for cynipids and nothing but decorative bumps for oaks, and that is exactly how the literature on galls depicts it. If that is the case, however, why hasn't natural selection enabled oaks to protect their own ecological interests with evolutionary responses that help mitigate cynipid attacks? I think it has, and this is how. Cynipids are herbivores that feed within plant tissues. Let's imagine for a moment that galls did not form around cynipid eggs and larvae. In that case, the larvae would not concentrate all their feeding in one location as they do within a gall; instead, they would tunnel through leaves or stems over a much wider area, very likely damaging oak vascular tissues as they went. With no physical constraints on their feeding, selection might have favored large individuals, larger than what can be achieved within galls, which would have, in turn, led to even more tissue damage to oaks from cynipids. Rather than thinking of galls as one-sided adaptations that favor only cynipids, we might more accurately view them as an evolutionary compromise that confines cynipid herbivory to one tiny site, minimizing damage from cynipid herbivores, and constraining the size of these oak parasites, while galls simultaneously allow cynipids to complete their development within the relative safety of a gall.

Dust on the wind

Gall initiation is not the only thing to watch for on your oaks in April. Depending on where you live, April is most often the month that oaks flower. Both male and female flowers are produced by the same tree, but it is the male flowers that are conspicuous. Shortly after cynipids have finished laying eggs in oak buds, you may notice the first growth of male flowers arranged lengthwise along spindly catkins, some 4 to 5 inches long, at the ends of branches throughout your trees. Most oaks will not begin to flower until they are at least 17 years old, so watch for the appearance of catkins on older trees. It is hard to miss oak catkins, but the female flowers of oaks are not only easy to miss, they are hard to find at all. Female oak flowers are tiny and nestled singly along thin branches higher up in the tree.

Most oaks are wind-pollinated. For best pollen transfer, catkins must lengthen, mature, and release their pollen quickly before the leaves fully expand and block the free flow of pollen in the air. Even though the female flowers of a particular tree may become covered with pollen from the catkins produced by that same tree, it usually takes pollen from a different tree to actually fertilize a female flower and initiate acorn development. And this is why it's important to have a population of oaks of each species in our

Oak catkins contain male flowers that release their pollen in April at our house.

landscapes rather than just single trees. Single specimen trees that do not have the opportunity to cross with other individuals of the same species will not produce acorns. As I write I am looking at a shingle oak I planted soon after we moved into our new home. It is the largest oak on our property, easily 60 feet tall. Every year thousands of tiny acorns form all over the tree, but because this is the only shingle oak we have and there are none nearby, all those nascent acorns, born of flowers that were not fertilized by a different tree, are aborted.

Unfortunately, oak pollen can trigger allergies in susceptible people. Although oak pollen is considered only mildly allergenic compared to pollen from many other trees, it can stay in the air fairly long under the right weather conditions and make some people truly miserable. Every time one of us sneezes at home, regardless of the time of year, Cindy and I joke that it must be oak pollen. The truth, though, is that at our house, oak pollen is only on the wind for a few days at the end of April. My hat goes off to people who are willing to tolerate a few days of discomfort each year in order to give their local environment the gift of an oak. Their allergenic effects aside, a full display of oak catkins gives an oak a pleasing ethereal appearance, even if only for a few days. Once pollen is shed, the catkins dry up and fall from the tree, adding important organic matter to the leaf litter below.

Polyphemus decorations

There is one more thing to look for on your oaks in early April: cocoons of the magnificent polyphemus moth, our second-largest giant silk moth, coming in just behind cecropia moths with a wing span of 6 inches. You could start your search any time from late September on, but April is the easiest time to find these cocoons because that's when the last of the marcescent leaves drop from the tree, exposing the cocoons like silvery white Easter eggs. Polyphemus cocoons are large as moth cocoons go, often reaching 2 inches in length, and are suspended from a twig by a silken rope another 2 inches long. But these days they are less and less common; if you find one on your oak tree, consider yourself lucky.

In most places, the cocoons overwintering on oaks are the product of the second generation of polyphemus moths. The pupa resting within the tightly woven silken cocoon deserves our respect, for it and the egg and caterpillar that created it, as well as the female moth that laid that egg, have all successfully run the natural world's gauntlet of predators, parasitoids, and diseases and a human-created gauntlet of deadly obstacles. After mating, mother polyphemus had to search for an appropriate host plant on which to lay her eggs. She confined her search to the hours after dark to avoid hungry birds, but that put her at risk from bats and owls, both of which prize giant silk moths as prey items. She needed to fly among the matrix of plants in our landscape to find the scent of my oak using her large plumose antennae

ABOVE
The odds of a
polyphemus
moth making
it to adulthood
are very small.

LEFT
Polyphemus
moths spin large
cocoons that
hang on oak
branches through-
out the winter.

and then zero in on it in order to lay an egg on a leaf. This was a critical stage in her reproduction; if she mistook my neighbor's Callery pear for an oak, larvae emerging from eggs after such a mistake would starve because they would not be able to digest the leaves of this non-native plant. And she had to do all of this in a landscape full of houses, garages, and barns bearing bright security lights that, for reasons not clearly understood, lure moths to their deaths like Sirens.

Female polyphemus moths carry about 250 eggs, but they cannot lay them all on the same tree, for if they did, chances are good all would be lost to natural enemies. When a predator finds and eats an egg or larva, it then searches all around the immediate area in case there are more nearby. For this reason, natural selection has favored females that disperse their eggs on host plants far and wide, over as large an area as possible. And polyphemus enemies are everywhere—ants, spiders, assassin bugs, damsel bugs, predaceous stink bugs, and particularly sphecid and vespid wasps, which hunt caterpillars day in and day out, as well as braconid and ichneumonid parasitoids, nuclear polyhedrosis viruses, many bacterial and fungal diseases, and legions of hungry birds. All these natural enemies attack caterpillars so relentlessly that just a few of the hundreds of eggs laid by the female will live to pupate. This is yet another reason to plant more than one oak on your property if you can; it will give local polyphemus moths the option of avoiding as many enemies as possible.

May

I'M NOT VERY GOOD at foreign languages (okay, I'm terrible at them), nor am I good at deciphering bird songs. I'm convinced the two talents are related. Fortunately for me, though, Cindy is excellent at both. So when she excitedly whispers, "Magnolia in the front oak," I pay attention. In fact, I grab my camera, not because I think a tree from China with pretty pink flowers has snuck into our yard; instead, I know Cindy has heard one of our prettiest migrants, the magnolia warbler (*Setophaga magnolia*). During peak migration in May, Cindy doesn't whisper just about magnolia warblers; last spring within one 30-minute stretch, she alerted me to a northern parula, a black-throated green, a black-and-white, and a magnolia—all warblers foraging in our oak tree. We do not live on a major migratory flyway, but redstarts, yellow warblers, blue-gray gnatcatchers, white- and red-eyed vireos, hermit thrushes, indigo buntings, Canada warblers, Baltimore and orchard orioles, eastern kingbirds, ovenbirds, Kentucky warblers, common yellowthroats, and blackpoll warblers regularly stop in our yard while on their way to breeding grounds farther north. They stop for one primary reason: to eat.

Magnolia warblers are one of our prettiest migrants, stopping over on their journey north each spring to forage for insects in our oaks.

Justifying migration

There are 650 species of birds that breed in North America. More than half, some 350 species, in fact, are long-distance migrants—birds that spend up to seven months of each year in the tropics but fly thousands of miles north to reproduce. It may seem puzzling that they should make such an effort; the physiological strain on migrants is difficult to comprehend. Migrants lose as much as 35% of their body weight if their migration route takes them over the Atlantic Ocean or across the Gulf of Mexico, and many die of exhaustion before they make landfall (Kerlinger 2009). Once in North America they can fly 300 miles in a single night if they have a tailwind, but when they stop for the day to rest, they must also refuel. In spring, the energy that fuels migrations comes primarily from insects rich in fats and proteins. In fact, a migrant will increase its body weight 30 to 50% each day they spend in stopover sites by eating insects (Faaborg 2002), that is, if there are enough insects to eat. And then there are challenges from uncooperative weather conditions, particularly from storms and stalled fronts in the spring and hurricanes in the fall. In short, migration is the most dangerous and taxing thing a bird will do in its lifetime. Yet, as with any other life history trait, the ecological benefits of migrating north during the spring to breed, and then back south to the tropics for the fall and winter must have outweighed the costs, or the behavior would not have evolved in any bird species let alone in hundreds of them. And indeed, when migration evolved, the benefits *did* outweigh the costs; birds that flew north to rear young could raise more offspring than birds that didn't.

Ever since the most recent glaciers retreated and during every intergla-cial period before that, the temperate zone offered something to birds that the tropics did not: a nearly inexhaustible supply of insects. Each spring across North America there was an explosion of fresh, tender foliage that was followed closely by an explosion of the insects that ate that foliage—a resource bonanza for birds that reared their young on insects. According to the Cornell Lab of Ornithology, birds that migrated north to take advantage of this enormous pulse of food could raise four to six young each year, a much better reproductive output than the two or three off-spring their tropical relatives could muster, and apparently worth the dangers and stress of the migra-tion required to do so.

And so, the evolution of bird migration was triggered and sus-tained for millennia at least in part by the seasonal flush of insects in the temperate zone. Some orni-thologists hypothesize that lower predation rates in the temperate zone also played a role in the evolution of migration. Regard-less, migration could continue to

Black-throated blue warblers fly from Cuba to New England to breed because of the abundance of spring insects in the north.

be a viable strategy as long as migrants are able to balance mortality from risky migration with enhanced reproduction once they arrive in breeding grounds. This, however, still requires diverse and abundant insects everywhere birds need to breed, for it is those insects that migrants turn into baby birds. But therein lies the problem: wherever we have reduced the absolute number of plants in an area, or replaced the native plants that support insects with non-native plants that do not, insect populations, and thus the ability of migrants to balance their risk equation, have been devastated. Devastated is a strong word, but at the University of Delaware my students and I have been measuring the impact of invasive and ornamental non-native plants on insects and the birds that need them for over 15 years: we have the numbers to back this statement up.

The magnitude of the impact non-native plants have on caterpillars is illustrated well by a study Melissa Richard, Adam Mitchell, and I conducted recently (Richard et al. 2018). We measured what happens to caterpillars when non-native plants displace native plants in agricultural hedgerows. Finding hedgerows that were thoroughly invaded by introduced plants such as autumn olive, multiflora rose, Callery pear, porcelainberry, burningbush, and bush honeysuckle was easy. They typify the "natural" areas near the University of Delaware, where we did our study. The trick was finding hedgerows that were still relatively free of invasive plants. Using a combination of restored sites and areas not easily accessed by deer (which exacerbate the spread of invasive plants), we finally found what we were looking for: four invaded sites and four primarily native sites of similar size. Using replicated transects, we counted and weighed caterpillars at each site, once in June and again in late July. By every measure the caterpillar community was seriously diminished when introduced plants replaced native plants. Even though there was more plant biomass within the invaded hedgerows, there were 68% fewer caterpillar species, 91% fewer caterpillars, and 96% less caterpillar biomass than what we recorded in native hedgerows. To summarize these numbers in terms of the everyday needs of the animals that eat caterpillars, we found 96% less food available in the invaded habitats!

What do such reductions mean for caterpillar-eaters? It's not rocket science to assume that if there is less food available in a habitat, there will be fewer creatures that need that food. If you doubt that logic, fill your bird feeders and then count the number of birds that use those feeders for a day. The next day, repeat your bird counts, but put out only 4% of the seed your feeders can hold. Your birds will clean out your feeders in short order, and then they will leave. If they were depending on those feeders to rear young, their babies would starve, or, more likely, your birds would know there wasn't enough food to rear young successfully before they laid eggs, so they would not even try to reproduce in your yard. The problem, of course, is that most birds do not rear their young on seeds from our feeders or from any place else, or, for that matter, on fruits from local bushes; they rear their young on insects and spiders (that themselves needed insects to become spiders). And the overwhelming majority of the insects that become bird food will not be in our yards unless we have the native plant species on which those insects depend for growth and reproduction.

As logical as it is to assume that birds need adequate amounts of food to sustain their populations, we no longer have to make this assumption; my student Desiree Narango for the first time has measured the impact of non-native plants on bird reproduction in suburban yards (Narango et al. 2017, 2018). Using Carolina chickadees as model birds that do their best to breed in human-dominated landscapes, and our nation's capital as a stereotypical suburban habitat, Desiree found that non-native ornamental plants supported 75% less caterpillar biomass than native ornamentals. How did this loss of bird food impact chickadees? It clobbered them. When yards

The white-eyed vireo is just one of hundreds of bird species that would be unable to reproduce without an abundance of insects to feed their young.

contained more than 30% non-native plants by biomass (the average percentage of non-natives in the yards she studied was 56%), chickadees were 60% less likely to breed at all, and if they did try, they were unable to produce enough young to sustain their population. The good news from her study is that we now have a target to shoot for when designing our landscapes. Desiree's study suggests that we can support breeding bird populations indefinitely as long as at least 70% of the plants in our yards are productive native species. The next step is to test whether the composition of native plant species in our yards makes a difference in this regard. For example, will yards with native oaks, cherries, willows, and birches support more birds than yards with spicebush, ironwood, black gum, and Kentucky coffee trees? The latter are all native species, but they differ markedly in the number of caterpillar species that use them. Our prediction is that yards with plant species that support oodles of caterpillars will also support more breeding birds, but we need to measure it to be sure.

Any birder worth her salt already knows where to look for spring migrants: look to the oaks! Can millions of birders be consistently wrong? I doubt it. But again, it never hurts to measure such things. And that is what Christy Beal, another of my students, did a few years back. Controlling for tree size, Christy compared the number of minutes warblers foraged among 15 different genera of trees during spring migration in New Jersey. Warblers foraged three times longer in oaks than in pines, the next closest competitor, and six times longer in oaks than in birches; they spent very little time foraging in any of the other 12 tree genera. Keep in mind that birds don't care two whits which tree they forage in as long as there is food there. They do, however, care about foraging efficiently. They cannot afford to waste time and energy searching for food where it doesn't exist, so they stay in unproductive trees as long as you would shop in Acme if all the shelves were bare—only a few seconds.

My point here is simple: Cindy and I enjoy the passage of migrants and the frenetic but hopeful work of breeding birds in our yard because we have trees like oaks that produce the insect food these birds need—and you can too!

Inchworms in the sky

In most of the country, May is the first month that you can walk into your yard and prove to yourself that oaks are great caterpillar producers. Fair warning: May is also the month that you will encounter the stiffest competition when hunting caterpillars on your oaks because that's when the most migrants stop in your yard to refuel on caterpillars and when resident birds have started to breed, feeding hundreds of caterpillars each day to their growing chicks. I often get puzzled looks when I talk about certain plants "making" caterpillars, but if you think about it, that pretty well describes what is happening. Caterpillars are built from the energy and nutrients stored in the leaves they have eaten. In effect, caterpillars are repurposed leaves that can walk. Plants that have developed a coevolutionary relationship with lots of species of moths and butterflies "make" or "produce" more caterpillars than plants that have not.

Caterpillars are really just walking leaves, for that is all they eat.

When I was young my family camped at a lake in North Jersey each summer. We would set our tents up in late May and use them repeatedly throughout the summer before we dismantled the campsite in September. Back then the tents were made of heavy canvas, and setting them up was no trivial task, particularly on a hot day. Although my father was a patient man, tent-raising day often tested his limits. The proverbial straw that broke my father's back on more than one such occasion was the steady rain of green caterpillars parachuting on silken threads out of the large oaks that shaded our tent site. For some reason, he didn't enjoy it when one would land on his nose or float down inside the back of his T-shirt. In my father's eyes, this flush of spring caterpillars was simply a nuisance that we would all have been better off without. If my father had been a bird, though, he would have viewed this bounty of caterpillars differently; he would have understood that these caterpillars were not only breakfast, lunch, and dinner, they were the key to successful reproduction and exactly what made that patch of New Jersey forest a Mecca for breeding birds.

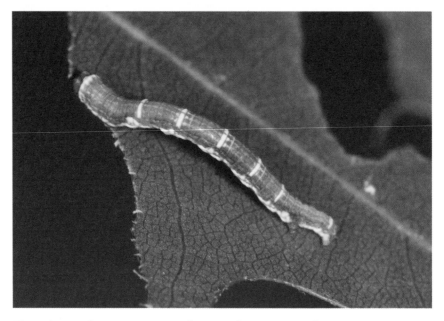

This one-spotted variant is just one of many inchworm species that are common on spring oak foliage.

Fortunately for birds, the caterpillars they like most—the smooth-skinned, tasty inchworms in the family Geometridae—are (or at least used to be) plentiful in the spring. Species like the porcelain gray, the fringed looper, the one-spotted variant, and the fall cankerworm (so named because the adults emerge and lay their eggs in the fall) can be so abundant on oaks that they occasionally trigger a homeowner's kneejerk reaction to spray, purportedly in order to "save" the tree, but I suspect the real reason is that many of us don't like little green "worms" parachuting into our hair. Be that as it may, spraying your caterpillars sets in motion a chain reaction that ends in local ecological disaster. When we spray a tree to kill caterpillars, we first kill all natural enemies of those caterpillars, which are more susceptible to our insecticides than the target caterpillars. Spraying rarely kills all the caterpillars, so the following year, there is an explosion of caterpillars because there are not enough predators and parasitoids around to keep their numbers down. So we spray again. The predators that eat caterpillars include local birds that were unable to raise young the previous year because we had killed the caterpillars they feed their young. Consequently, each year we spray, there are fewer birds around to control caterpillar populations, and so we spray again. Soon, we are locked into spraying toxins all over our properties every year because we have eliminated the natural controls that kept those caterpillars in check—all this to save trees that are well adapted to spring caterpillar flushes and, unless otherwise stressed, would not have been harmed by the caterpillars in the first place.

Oak underwings

If you are lucky, another type of caterpillar you might encounter on your oaks in May is one of the underwings (*Catocala* spp.), so named because the second pair of wings on the adults are usually brightly striped in black and red, yellow, or orange. Though these are rather large caterpillars when full grown, reaching a few inches in length, you will have to look carefully to find them. They feed on oak leaves at night, but during the day they crawl a good distance from their feeding site and sit motionless on the branches or trunk of the tree itself. That presents a formidable challenge for underwing hunters, for most species look exactly like the bark they are sitting on. The easiest way to find underwing caterpillars is to cut a swath of burlap about 12 inches wide and long enough so that it can encircle the trunk of your oak. Tie the burlap around the oak with string so that 6 inches of burlap hang down on either side of the string. When older underwing caterpillars crawl down the trunk to rest during the day, they will often crawl up under the burlap flaps where they are easy to see.

Nationwide, there are at least 17 species of underwings that use oaks as their sole host plant. Carl Linnaeus, the father of modern taxonomy and the first to consistently use binomial nomenclature (genus and specific epithet) when describing a species, adopted a female/marriage theme for the common names of underwings. Later taxonomists followed his lead, resulting in oak underwings with such names as the little nymph, the girlfriend, the bride, the connubial, the little wife, the widow, and the Delilah. Underwing larvae complete their growth in late spring, pupate underground, and emerge as adults later in the summer. Eggs are then laid in bark crevices, where they spend the entire winter before hatching in early spring.

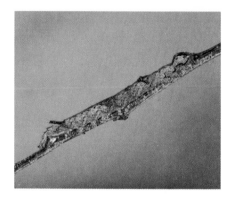

Though large when full grown, underwing caterpillars blend in well with oak bark.

Oak jewels

If you are lucky enough to live in the Southwest, you might encounter the only U.S. representative of the moth family Dalceridae. Dalcerids, or jewel caterpillars, do indeed look like beautiful, precision jewels, often in dazzling reds, greens, and yellows. Our one species, *Dalcerides ingenita*, will not win a contest of the prettiest jewel caterpillar on earth, but it is perhaps our most unusual oak-feeder and certainly pretty enough. There are two

Our only jewel caterpillar, *Dalcerides ingenita*, is unique among North American oak moths.

generations per year in Arizona, so you could find *D. ingenita* caterpillars on Mexican blue oak, Arizona white oak, or Emory oak in May and again in August, or perhaps the solid orange adult moth at your front porch light in June or September. If you do, it will be easy to appreciate one of nature's most unusual creations.

Thaxter's sallow

Early in April, one of the first moths to emerge from winter's sleep, regardless of temperature, is Thaxter's sallow (*Psaphida thaxteriana*), named after Roland Thaxter (1858–1932), a prominent lepidopterist from Maine; but the eggs that females scatter on oak twigs do not hatch until the tree's leaves are nearly fully extended in May. Thaxter's sallow is just that, colored in mottled grays that enable it to disappear in plain sight when it sits on oak bark. From the start, Thaxter's sallow caterpillars are secretive loners. Though they have a row of white triangles along their dorsum that stand out against a dark ground color, the caterpillars do not advertise this striking pattern; instead, they hide within a cluster of young oak leaves they have tied together with silk. Why natural selection has favored such a distinct color pattern only to hide it away within leaves is still a mystery.

Like so many other caterpillar species, Thaxter's sallow is uncommon and becoming more so each year. In fact, it is now a species of conservation concern in several eastern states. It is difficult to prove what the exact cause of this decline is, but if I were in charge of its salvation, I would look first to the plants on which it depends for its existence. Thaxter's sallow is an oak specialist and seems to have an affinity for white oaks and scarlet oaks throughout much of its range. I do not think it's a coincidence that the decline of this beautiful caterpillar correlates well with the rapid suburban sprawl that has destroyed oak forests in the eastern United States. But if everyone were to replace their Callery pear—a highly invasive ornamental tree from Asia that is still promoted by the nursery trade in too many places (particularly as the cultivar 'Bradford')—with an oak species that does well in their bioregion, my guess is that Thaxter's sallow would once again become common enough to help keep our neotropical migrants well fed.

The destroyer

We can't talk about caterpillars on oaks in May without mentioning the notorious gypsy moth. The gypsy moth was accidentally on purpose introduced from Europe to Massachusetts by E. Leopold Trouvelot in 1869, believe it or not, because of poor taxonomy. Trouvelot was out to make a better silk moth. His hope was to cross gypsy moth (*Lymantria dispar*) with the commercial silk moth (*Bombyx mori*), to introduce the vigorous genes so prevalent in gypsy moths into the inbred commercial silk moths and thus create a beast that was healthier and made more silk. The scheme wasn't quite as harebrained as it may seem because in Trouvelot's time, the

The destroyer, *Lymantria dispar*, has proven to be one of our worst invasive species.

gypsy moth was mistakenly placed in the genus *Bombyx*. But even if this taxonomy had been accurate, the idea of crossing gypsy moths with silk moths would probably still have failed because most species won't mate with other species, even if they are in the same genus. If they did actually mate, chances are good their offspring wouldn't be viable. We now know, however, that the gypsy moth not only evolved within a different genus, it evolved within the Erebidae, an entirely different family from the silk moth's family, Bombycidae. The gypsy moth and silk moth were no more likely to successfully pair than a dog and a cat!

But Trouvelot didn't know this, so he proceeded with his plan. He brought a colony of gypsy moths from Europe to his home in Medford, Massachusetts, built them a nice sturdy cage, and watched in dismay as a storm blew the cage to pieces, releasing his pets into local oak woods.

Within 10 years, *Lymantria dispar* (from the Latin for "destroyer") was defoliating New England and well on its way to becoming the worst forest pest in North American history. Gypsy moths continue to expand their range, killing oaks as they go. In fact, if my father were putting up tents in North Jersey today, he would not, as in the old days, be bothered by inchworms parachuting from the oak trees above. The gypsy moth swept through those woods of my childhood in 1975, killing so many oaks that their numbers and those of the inchworms that are so important to bird migration and breeding are now a fraction of their pre–gypsy moth splendor.

You might wonder why I speak of inchworms and other native caterpillars almost with reverence but disparage introduced species like the gypsy moth. They are all caterpillars and aren't all caterpillars the same? In one sense they are, but the ecological impact any one species has, be it positive or negative, depends on where you are. If I were in Europe and Asia, I would not speak ill of the gypsy moth. It evolved in concert with the plants and other animals in Eurasia, and though it is an outbreak species that occasionally defoliates on a small scale in Eurasia, it is not "the destroyer" there that it is here—for one simple reason: it has not been removed from the suite of natural enemies that keep its explosive potential in check. There are over 100 species of parasitoids that attack gypsy moth in Eurasia, as well as several specialized beetle predators and other invertebrate predators, not to mention important diseases that, in concert, usually maintain gypsy moth populations below problem levels. Not so here in North America; when Trouvelot brought his gypsy moths to Medford, he brought them alone, leaving their natural enemy complex behind. Not even our hungry birds pitched in; as a rule, birds eschew hairy caterpillars like the gypsy moth. This North American imbalance is not unique to gypsy moth; the same phenomenon occurs with satin moth, winter moth, emerald ash borer, hemlock woolly adelgid, Asian long-horned beetle, Japanese beetle, and viburnum leaf beetle, among others, as well as introduced plant species. Species that are moved around the world without their natural enemies often experience what ecologists call "enemy release"; with no predators or diseases to slow their population growth, they often cause enormous ecological upheaval in their new lands.

Understanding oak leaf shape

In most areas of the country, leaves on deciduous oak species have expanded to their full size and shape by mid-May. If I were to assemble fully expanded leaves from all our North American oaks into one photograph, it would be an impressive display of biological variation. Many species, of course, have leaves that look just like stereotypical oak leaves. But the leaves of other species look far more like willow leaves than oak leaves, and several species from the Southwest and California have leaves that are spitting images of holly leaves. Some species have big leaves, some have small leaves. Some leaves have lobes with pointy edges, some have rounded lobes. Even individual trees produce leaves that vary a great deal in size and shape. What is all this leaf variation about?

A small sampling of the variation in leaf size and shape found in North American oak species.

Leaf shape tells us both evolutionary and ecological stories. Evolution-ary stories are usually more of a challenge to read than ecological stories, and it is difficult to test whether we have read the stories correctly or not. Certainly, some degree of leaf variability has come about through the pro-cess of speciation itself. Tree populations that become isolated from each other, usually due to a geographical barrier like a mountain range or a large river, necessarily stop exchanging genes and over immense periods of time evolve into distinct species that no longer look alike. The evolutionary changes that accrue over time can result in modifications to the original leaf shape, and these modifications may be caused by random genetic changes that do not reflect specific functions as well as by peculiarities of the new environment. Ecological function, however, is a powerful shaper of leaves that is easier to interpret than evolutionary history if we clearly understand what job leaves have been assigned by natural selection.

The primary function of plant leaves is to gather energy from the sun and convert it through photosynthesis to simple sugars and carbohydrates. But, as we all know, sun energy creates heat. Leaves can tolerate some heat from the sun, but too much of that heat disrupts the chemical processes occurring within the leaf. Now hold that thought. To build sugars and carbohydrates, leaves need carbon, which they get from airborne carbon dioxide. For the chemical reactions that comprise photosynthesis to take place within leaf cells, leaves must draw carbon dioxide in through small openings in their surfaces called stomata. Unfortunately, an open stoma not only permits carbon dioxide to enter a leaf, but it also allows moisture to escape from the leaf. The daily life of a leaf, then, is a delicate balancing act between taking in enough carbon dioxide for photosynthesis to proceed without losing so much water that the leaf wilts. And it must maintain this balance without overheating. A leaf does this by constantly opening and closing its stomata throughout the day, and through its size and shape.

You may have noticed that the leaves on the lower branches of an oak tree are often much larger than the leaves at the top of the tree. There's a reason for this. The leaves at the bottom of the tree are shaded by the leaves above them. In order to capture enough sunlight to fuel photosynthesis, lower leaves become large and broad and usually have few lobes. Leaves at the top of the tree, in contrast, are exposed to so much sunlight that they do not need to be very large to acquire all the energy they need for photosynthesis; instead, their main problem is overheating. Deeply cut lobes and smaller size, limiting their exposure to the sun, are two ways treetop leaves reduce leaf area and thus their heat problem.

These same conflicting forces dictate leaf shape in challenging environments like the hot, dry regions of the Southwest. To conserve water and minimize heat buildup, oaks in dry regions typically have small leaves throughout the tree, and each leaf is comparatively thicker than oak leaves in cooler climes. They also are endowed with ample coatings of wax on their surfaces to help minimize water loss. Moreover, many western oak species are shorter in stature than some of the giants in the East and would have been within easy reach of the large Pleistocene mammals that were such an important ecological force in the recent past—thus, the evolution of holly-like spines that adorn the leaf edges in these trees. And finally, let's not forget that there often is more than one way to skin an ecological cat. That is, there often is more than one solution to reaching an evolutionary compromise to the aforementioned physiological constraints imposed on leaves. For example, willow oaks (*Quercus phellos*) and water oaks (*Q. nigra*) often grow in the same southern habitats, yet their leaves have very different shapes. Willow oak leaves are small and narrow; water oak leaves are larger and flare at the terminal end. You might think willow oak's growth rate would suffer due to less leaf area, but willow oaks compensate for their small leaves by producing many more leaves than water oaks. The result, if anything, is that willow oaks have more leaf surface area than water oaks and they grow equally fast.

June

ONE DAY IN mid-June 1987, while I was driving down I-95 to a research site in Maryland, I was startled by a substantial thwap on the car's windshield. What was that? The smear on the glass suggested it had been a sizable insect, but it was too early in the season to be a praying mantid, katydid, or mature grasshopper. Thwap! Thwap! Suddenly, whatever was thwapping my car was everywhere, flying across the road, up in the air, down to the ground, hither and yon. Too curious to continue on, I pulled the car over, got out, and immediately saw that I was in the midst of a 17-year cicada emergence. Oh, wow! I hadn't seen one since 6th grade when I lived in New Jersey.

Periodical cicadas

Periodical cicadas put on one of nature's most impressive displays of synchrony and slow development. Sometimes mistakenly called 17- or 13-year locusts (true locusts are a type of grasshopper), cicadas of all types are distant relatives of aphids, treehoppers, and other hemipterans. Rather than chewing their food with mandibles, they suck it up with mouthparts shaped like short straws. Sucking mouthparts work well for lots of insects including cicadas, but it is hard to applaud their choice of what they suck as immature nymphs: the xylem in tree roots. Though plentiful, xylem is just this side of water in its nutritional value. It contains almost no nitrogen and very little in the way of carbohydrates. It also requires powerful sucking muscles to draw it out of a plant. All this is in stark contrast to plant phloem, which carries more nutrients than xylem and is under positive pressure; if an insect like an aphid plugs into a phloem tube, it doesn't have to suck at all. The phloem juices flow out all on their own. And this is one of the reasons it takes periodical cicadas more than a decade to reach maturity; by volume the vast majority of their food is water and they have to work hard to get it!

There are seven species of periodical cicadas in the United States (three species that emerge every 17 years and four that emerge every 13 years), and they are separated into 23 broods united by the synchrony of their emergence, geography, and the time they require to mature. Some areas of the country (Maryland, for example) have both 13- and 17-year broods, and it is hard to keep track of which brood will emerge next; but generally our northern states, ranging as far west as Nebraska, have 17-year cicadas, while the Deep South has only 13-year broods. When periodical cicadas emerge, they do so by the millions and all at once. We call them periodical cicadas to distinguish them from the more than 150 species of annual cicadas in North America, which appear as adults every year in mid-summer. Their common name and the fact that we see annual cicadas every year makes it seem like they have a one-year life cycle. Not so. Like periodical cicadas, annual cicadas spend years developing underground, on roots; but they differ from periodical cicadas in that they do not emerge in synchronous broods. Instead, some individuals in each population emerge every year.

Besides the enormity and suddenness of their appearances, periodical cicadas have another feature that is hard to forget: their loud and near continuous buzzing. Cicadas of all types communicate with tymbals, a set of ridges on each side of their abdomens. Strong muscles on either side of a tymbal buckle and unbuckle these ridges, with each buckle making a click by a mechanism similar to the clicking of a soda can by squeezing and relaxing your fingers. But unlike a squeezed soda can, tymbal muscles buckle and unbuckle the ridges so fast, 300 to 400 times per second, that the clicks blend into a loud buzz. This buzz is amplified by a concave resonance chamber located just below each tymbal. The end result is one of the loudest sounds produced by any animal. I have been in Costa Rican forests where the buzz of cicadas was so loud and at such a pitch that I had to walk with my hands over my ears. Only male cicadas buzz, and they do it, not surprisingly, to attract the attention of females. If a female hears a male nearby, and if she is receptive and likes the sound he is making, she will snap her wings. The male hears the snap and then moves closer to the female. After several bouts of buzzing and snapping, the pair is united and they mate.

Even though young cicadas (nymphs) develop underground, adult females insert their eggs aboveground in rows along thin live twigs of a host tree using a stout ovipositor strong enough to slice open the twig's bark. Once inside the twig, the eggs absorb water and mature. After six weeks they hatch, and the tiny cicadas fall to the ground, where they work their way into the soil until they find a tree root. Once there they insert their mouthparts and suck xylem until they are fully grown. Under good conditions, a mature tree can host 20,000 to 30,000 cicada nymphs on their roots with no apparent ill effects. After the time period appropriate for their species, all the nymphs crawl to the surface and molt into the adult stage within one or two days, a truly amazing display of synchronization after spending years underground. How they synchronize is still somewhat of a mystery; cues to emerge may include the temperature of the soil or something within the xylem itself. Whatever the cue, when they crawl from the ground the cicada nymphs leave an exit hole behind, and after a mass emergence it looks like someone drilled thousands of half-inch holes in the soil under every host tree. What also remains behind are the exuviae (shed skins) of all the cicadas. Usually the nymphs crawl a few feet up a tree trunk to shed, but if there is a shortage of tree trunks, they will just molt on the ground. Cicada exuviae are made of chitin, a very durable material, and they can last many months as a reminder of this special entomological event.

Why stay underground for 13 or 17 years? I already suggested that xylem is such a poor source of nutrition that cicadas need years to complete their development. But there must be some other reason; for one thing, there is good evidence that some species of large annual cicadas can complete their development underground in as few as four years. So, what factors have selected for the extra time delay? Ecologists agree that this curious and extreme life cycle is best explained by predator satiation, the very same phenomenon that helps explain periodic acorn masts. By remaining underground out of harm's way for 13 or 17 years, periodical cicadas reduce predation in two ways. First, mass, synchronized emergence overwhelms predator populations by sheer numbers. There simply are not enough squirrels, birds, opossums, raccoons, foxes, and other predators to eat all the available cicadas in a large emergence. Therefore, many, if not most, cicadas survive to lay their eggs. Second, such long periods of time between emergences prevents any one predator from becoming a specialist on periodical cicadas the way cicada killers (large sphecid wasps) have specialized on annual cicadas. If a wasp like a cicada killer "decided" (in the evolutionary sense) to attack only periodical cicadas, it would have to go 13 or 17 years between generations, a trade-off that even periodic "all you can eat" buffets does not justify.

The last time periodical cicadas emerged at our house was 2004, when my oak tree was only four years old. They are scheduled to emerge again in 2021. Whether my oak was large enough in 2004 to attract ovipositing female cicadas remains to be seen, but it is certainly large enough now. As with so many other types of insects, oaks are favorite host trees for periodical cicadas, and I am expecting quite a buzz in our front yard in 2021!

Oak treehoppers

Another group of fascinating creatures inhabiting my oak tree in June are inconspicuous compared to cicadas, even though they are close relatives. In fact, you have to purposely search for them as they suck plant juices from branch tips to even know they are present. These are the oak treehoppers, several species of *Smilia*, *Atymna*, *Microcentrus*, and *Platycotis* that specialize on oaks. All these "hoppers" have powerful hind legs that can launch their little bodies a distance many times their body length if they are disturbed. Some people also know them as thorn bugs: the top of their first thoracic segment, their pronotal shield, is usually expanded into bizarre shapes, some of which look just like rose thorns. The pronotal shield on oak treehoppers is modest compared to many other species but still distinguishes them from all other insects. No one really knows why treehoppers have pronotal shields. Some possibilities include defense (some shields have one to several spikes, which surely makes it difficult for birds or lizards to swallow them), crypsis (those that look like thorns are hard to spot on plant stems with real thorns), and mimicry (the shields on a number of species resemble distasteful, stinging ants and thus may be avoided by predators).

Treehoppers are very interesting in their appearance but are less so to watch; most of the time they just sit and suck. Some species do stand out, though, in how they reproduce and even resemble *Homo sapiens* in this regard more than most other insects. Rather than abandoning their eggs after laying them, females of the oak treehopper *Platycotis vittata*, for

Smilia camelus, one of the oak treehoppers, is more easily seen at your porch light than on oaks, where they blend in with their background.

example, stand guard not only until the eggs hatch but until their young-sters mature as adults. Extended parental care is rare in animals, and particularly rare among insects, in part because it is such a costly form of reproduction. By staying and protecting one group of offspring, mother platycotis sacrifices her chance to produce additional offspring. This stands in stark contrast to most other insects, and even to most treehoppers; rather than investing heavily in one batch of offspring, most species spread their reproduction over time and space. That is, they don't put all their eggs in one basket; instead they lay many batches, each in a different place, and then abandon their eggs immediately to go off and lay more eggs. In terms of offspring survival, this strategy works on average just as well as guarding; predators and parasitoids will find some egg batches but not all, and because they are not stuck guarding their first clutch, females that don't guard their eggs can lay many more eggs than those that do—and often create more surviving offspring (Tallamy 1999).

Platycotis vittata is one of the few insects that guards its young until they reach adulthood.

So why has *Platycotis* adopted maternal care as a reproductive strategy if there is a better way? Good question, and one that took me years to answer (Tallamy and Brown 1999). If you compare traits that are common to insects that have maternal care versus traits shared by species that do not, a striking pattern emerges. With few exceptions, insects that have adopted maternal care as a reproductive strategy are semelparous; that is, they lay only one clutch of eggs in their lifetime, whereas insects that do not care for young are iteroparous and lay many clutches before they die. I believe this is the key that explains when maternal care is an evolutionary option and when it is not. The ecological cost of maternal care is in lost opportunities for reproduction, but if there are no future reproductive opportunities, say, for example, because there is not enough time left in the season, or because the food needed to rear young is seasonally ephemeral, then there is no cost to guarding young. In that case, guarding young from enemies instead of laying many small clutches and abandoning them makes sense. When there are no seasonal or resource constraints, however, maternal care is too costly to ever take hold.

Platycotis vittata provides a perfect example of this phenomenon. *Platycotis* has two generations per year: one in the spring and one in the fall. Nymphs are restricted to spring or fall development because that is the only time nutrients are mobilized in the oak's vascular system. In the spring, females that have overwintered lay eggs in oak twigs much like their cicada cousins, but they do it right before bud break. The eggs hatch just as nutrients are being pumped up from the roots to the growing shoot tips. This timing places *Platycotis* nymphs on oak twigs when they can best intercept those nutrients and use them to grow quickly to adulthood. And none too soon, because after the initial flush of oak leaves, nutrients no longer flow in the tree's vascular system but instead are stored in the leaves until fall. This means reproduction is not an option during the entire summer! *Platycotis* adults produced during the spring generation spend the rest of the summer waiting to lay eggs just as nutrients flow once again, when leaves give up their nutrients for transport to the roots during the fall for winter storage. And so, iteroparous reproduction is not a viable option for *P. vittata*: in both the spring and fall generations, nutrients are not available for nymphal development long enough for females to produce more than one brood of young. Thus, it makes sense to spend time and energy protecting their first brood rather than abandoning them to produce additional broods.

Most people encounter adult oak treehoppers only at porch lights at night, and they are never very common. But you can look for *Platycotis* families lined up along oak twigs in both the spring and the fall: two opportunities to enjoy their beauty and ponder the interplay between treehoppers, nutrient transport within your oak, and treehopper natural enemies.

Caterpillars in June

In most places, the number of caterpillars using oaks is at its nadir in June, not because species avoid oaks in June but because breeding birds have already eaten most of them. The caterpillars that remain are either distasteful, too small for birds to bother with, hidden in tightly curled leaf edges or leaf mines, or are so cryptic they have escaped forging birds. An amazing

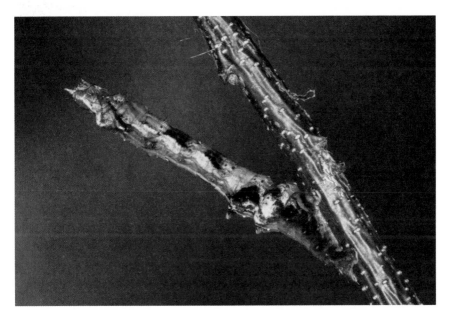

Kent's geometer (*Selenia kentaria*) often escapes bird predation because it resembles a twig so closely.

example of the latter is Kent's geometer (*Selenia kentaria*). From eastern Kansas to the Atlantic and Canada to Georgia, Kent's geometer is such a convincing twig mimic that it regularly escapes bird predation as well as curious humans. The best way to see if Kent's geometer is present on your oaks is to use a beating sheet (see sidebar) They are never abundant, but seeing a late instar larva is real eye candy for a naturalist. From their posture on a branch, to their background coloration, to the lichen-like scales encircling their posterior third, Kent's geometers are the quintessential twig mimics and a marvel to behold.

An equally bizarre inhabitant of your oak is the filament bearer (*Nematocampa resistaria*). If this caterpillar survives bird predation, its main enemies are tiny parasitoid wasps that constantly try to land on its back and inject one or more eggs. The filament bearer is aptly named because it uses four expandable extensions of its dorsal exoskeleton as Hydra-like bats, filaments if you will, that serve to knock parasitoids away

The filament bearer (*Nematocampa resistaria*) protects itself from tiny parasitoid wasps with four inflatable bats.

before they get a chance to insert an egg into its body. The filaments are hollow and can be retracted or rapidly inflated by withdrawing or injecting hemolymph (caterpillar blood) into them. Touch a filament bearer gently, and it will be happy to demonstrate this unusual adaptation.

Two heads are better than one

Although oaks support the development of 897 species of moths in the United States that we know of, for some reason only 33 species of butterflies and skippers use oaks as a larval host plant. Fifteen of these, however, are hairstreaks in the family Lycaenidae. The caterpillars of oak specialists like the white M hairstreak (*Parrhasius m-album*), the California hairstreak

COLLECTING CATERPILLARS

A common way to collect caterpillars is with a beating sheet. Just hold the sheet under the leaves you wish to collect from and tap the branch lightly with a stick. Caterpillars fall to the sheet and can then be examined, photographed, or collected. You do not need to whack the branch; whacking injures the bark and besides, what dislodges caterpillars is surprise, not force. An unsuspecting caterpillar can be jarred loose from the branch if it is caught off guard, but once it senses a disturbance, it clings tightly and is very difficult to pry off.

Beating sheet, in action.

(*Satyrium californica*), and the banded hairstreak (*S. calanus*) don't look much like caterpillars at all. Like all lycaenids, they look more like segmented, velvety slugs with no clear front or back. They do have a front, of course, but because their head is tucked down and hidden by their first thoracic segment, their front end looks just like the back end from above. Hairstreak caterpillars move very slowly, eat very slowly, and usually blend in with their background quite well.

As adults, however, hairstreaks are more interesting; they are small, well-behaved butterflies that often sit quietly while nectaring on flowerheads or just resting on leaves. It would be nice to think these butterflies sit still so we can get close with our cameras, but I'll bet there is another reason. Like many other butterflies, hairstreaks hold their wings straight up over their bodies, but unlike most other butterflies, they keep them still and do not constantly raise and lower them or flutter in any way. When you look at a resting hairstreak, then, you are looking at the underside of their wings which typically have only one splotch of color, usually orange or red, on the lower portion of the back edge of each hindwing. Curiously, there is also a distinct black spot, and each hindwing has a thin projection that, taken together, look all the world like a pair of antennae. And that, apparently, is the idea. Combined, these features resemble a butterfly head with eyes and antennae.

For over a hundred years, the false head of hairstreaks was explained as being an adaptation to draw the attention of birds away from the real head of the butterfly. Birds usually try to attack the head of an insect to make a quick kill, so if natural selection made birds think the head was actually at the posterior end of the butterfly, a bird could attack it and only get a mouthful of wing while the hairstreak escaped. This explanation seemed logical enough, but there were parts of it that didn't add up. First, hairstreaks are small insects and would not make much of a meal for most birds. How often did birds actually chase hairstreaks anyway? Second, the false head on the hindwings is even smaller than the butterfly; could a bird really see it well enough to be misdirected in its attack? Finally, if a bird (or human with an insect net) did chase a hairstreak, the butterfly always flew directly toward the sun, which temporarily blinded the predator if it persisted in the chase. Flying into the sun undoubtedly works well; why, then, bother with a false head?

In 2013 Andrei Sourakov published a paper that resolved these questions and set the record straight on hairstreak behavior. He noticed that when hairstreaks are resting on leaves or flowerheads, they are not as motionless as they seem; the false antennae on their hindwings are constantly and alternately moving up and down, just as if they were real antennae. These movements are certainly too subtle to have evolved as a defense against bird predation, but they are perfect for catching the attention of jumping spiders (Salticidae). Salticids hunt just as their common name suggests: they have excellent eyes (eight of them, in fact) and when they spot an insect, they jump on it. Salticids are fantastic jumpers, launching themselves up to 50 times their body length toward their prey, usually from so far away, the victim doesn't know the spider is near. And like birds, they always jump toward the victim's head. Salticids often hunt prey several times larger than themselves, so they have to make a quick kill or the stronger prey item

Like all hairstreaks, the red-banded hairstreak (*Calycopis cecrops*) displays a false head on its hindwings.

would escape. With a series of simple experiments, Sourakov showed that the false head of a hairstreak is irresistible to a hungry salticid. The spiders always jumped on the wrong end of the butterflies. More often than not, the false head on the wings would then break off, allowing the butterfly to escape, a little marred but alive. The ability to protect yourself from unseen enemies is important if you are a hairstreak because salticids are some of the most abundant predators in vegetation!

Most hairstreaks that develop on oaks do so in the typical way: their caterpillars eat green leaves. But a few, such as the red-banded hairstreak (*Calycopis cecrops*), have chosen a different route: rather than eating living oak leaves, they spend their caterpillar life stage in leaf litter below oaks, eating dead oak leaves. We can only guess what selective forces favored this odd departure from the norm, for having sent most of their nitrogen to tree roots before they senesce, dead leaves are notoriously poor in nutrition. Perhaps there are fewer predators in leaf litter, although that is hard to imagine. But the red-banded hairstreak is not the only lepidopteran that develops on leaf litter; some 70 species of moths do so as well in the United States

Salticid jumping spiders like this beautiful species from Arizona are often fooled by the false head on hairstreak wings.

alone, so there must be a good reason for this odd larval diet. Red-banded hairstreaks are among our most common lycaenids, but I wouldn't spend a lot of time looking for their larvae; they are nigh on impossible to spot among thousands of brown oak leaves. Knowing they are there, however, gives us another reason to create safe sites for leaf litter and the abundant life it supports somewhere on our properties.

The most unusual of all the hairstreaks that depend on oaks is Edwards' hairstreak (*Satyrium edwardsii*). In appearance it is similar to other hairstreaks: grayish underwings with rows of dashed spots here and there. In fact, it can be a challenge telling one adult hairstreak from another in the field. Their larvae also look like other hairstreak caterpillars. But in behavior they differ markedly. Edwards' hairstreak spends the winter as an egg glued to oak twigs. Soon after bud break, the egg hatches and the tiny caterpillar begins to eat young leaves and oak catkins. As the caterpillar grows, it develops glands on its abdomen that secrete a sugary substance laced with amino acids, the invaluable building blocks of proteins. In combination, this secretion is highly attractive to ants and when they find a caterpillar producing this valuable food supplement, they protect it from predators and parasitoids as only ants can.

So far, Edwards' hairstreak is similar to many species of lycaenids that have a mutualistic relationship with ants. In exchange for sugar and amino acids, ants protect the caterpillars from predators. But when Edwards' hairstreak caterpillars reach the third instar (about half grown), they do something other hairstreaks do not: they climb off their host oak as dawn approaches into a protective byre, a den of sorts that their guardian ants have constructed out of leaf litter. There, they rest for the day, often in groups, and then climb back up into their host tree come evening, where they feed again on leaves through the night. During each of these moves the caterpillars are guarded closely by ants. And on it goes—down the tree at dawn and back up at dusk—until each caterpillar reaches its full size. At that point the hairstreaks make one final trip down the tree trunk into their byre, where they form their chrysalis. Ten days later they eclose as adults, expanding their wings and crawling out of their byre for the last time to begin their new life as a butterfly.

July

NOT LONG AGO I received an email from a homeowner who lived on the Eastern Shore of Maryland. She was distressed about the mistletoes within the branches of her oaks. "This invasive species is going to kill my oaks. What can I spray to get rid of them?" I started my reply by setting the record straight about the geographic origin of mistletoes. They are not invasive species (non-native species that are displacing native plant communities) but rather native species doing what they have always done to oaks. In most cases, this is very little, but during extreme and prolonged drought they can weaken and occasionally kill oaks if the tree has a heavy mistletoe load. Oak-inhabiting mistletoes (*Phoradendron* spp.) are hemiparasites, meaning they are just a little bit parasitic. They photosynthesize 98% of their energy with their own green leaves, but they do sink roots into host branches that penetrate the xylem and suck up some water. Studies on California mistletoe, a hemiparasite on several western oak species, showed that it had no ill effects on its host oaks and acted more like an epiphyte than a parasite (Koenig et al. 2018). And these aerial shrubs are contributing epiphytes at that! Mistletoes flower in the fall and produce berries that ripen near the end of winter. These berries are often the only berries available by winter's end, particularly for blue-birds. My advice with normal mistletoe densities on your oaks is to do nothing, except maybe kiss your sweetie if you can maneuver him or her under one.

Great purple hairstreak

If you tolerate some mistletoes on your oaks, and if you live in any of the southern states across the country, you may be lucky enough to have a population of the great purple hairstreak (*Atlides halesus*), without a doubt our largest and most spectacular hairstreak. Like so many lepidopterans, they are host plant specialists, in this case developing only on mistletoe leaves. The great purple hairstreak is actually a tropical species that ranges as far south as Panama, but as mistletoe moved north during our last post-glacial period, the butterfly followed it into North America as well. In most places there are three generations each year, so you have ample opportunity to observe these beautiful butterflies. Males typically sit in treetops, hoping to spy and intercept females that are searching for mistletoe, but both sexes nectar at flowers like goldenrods (*Solidago* spp.), Hercules' club (*Zanthoxylum clava-herculis*), sweet pepperbush (*Clethra alnifolia*), and wild plum (*Prunus americana*), providing the best chances to see these wonders.

Throughout their range, great purple hairstreaks share their *Phoradendron* mistletoes with the beloved emarginea (*Emarginea percara*), a small moth beloved for perfectly mimicking the coloration of the lichens that typically adorn oak branches. You are not likely to find emarginea caterpillars by searching mistletoes, or adults by searching lichens; their crypsis is too good, and you will throw up your hands in defeat before long. Fortunately, though, the beloved one readily comes to lights at night. No matter how many times I encounter this beautiful moth, I cannot resist taking its picture yet again.

The exquisite great purple hairstreak (*Atlides halesus*) is a specialist on oak mistletoes.

The adults may be brilliantly colored, but the larvae of great purple hairstreaks blend in well with mistletoe leaves.

A stunning specialist on mistletoe, you can find the beloved emarginea (*Emarginea percara*) wherever mistletoes and lichens adorn oaks.

Yellow-vested moth

A number of caterpillars graze on the upper surface of oak leaves, removing the parenchymal cell tissue in between leaf veins as they eat. This feeding mode is called skeletonization because it leaves the "skeleton" of the leaf intact. But an upper leaf surface is a dangerously exposed place to be, so many species tie another leaf over the area they are grazing with silk, creating a shelter that hides their activity from hungry birds and spiders. Perhaps the handsomest caterpillar species that behaves this way is the yellow-vested moth (*Rectiostoma xanthobasis*). This dapper little guy is aptly named, for he looks all the world like a butler wearing a yellow vest. Look for him in mid-July wherever oaks grow, from the Ozarks eastward and as far north as Vermont.

The yellow-vested moth (*Rectiostoma xanthobasis*) is one of the showiest leaf skeletonizers on oaks.

Katydids

Once upon a time, there was a young woman named Katy who fell in love with a handsome young man. Alas, he did not share her feelings, and he married another. Soon thereafter, he and his young bride were found poisoned in their bed. Who perpetrated the crime? That was never determined, but some say the insects in the trees were watching that night, and each summer they solve the mystery by singing "Katy did, Katy did!" Or so the legend goes.

From age four through my early 20s, I was lucky enough to spend a good part of every summer camping in North Jersey with my family. A milestone of each summer, marking the midpoint of our camping adventures, was when various species of katydids—large, grasshopper-like insects—began their nightly chorus in mid-July. Not only did the katydids tell us that summer was half over, but they also signaled the approximate time each night. The katydids in the oaks above our tents were loudest before midnight

The *Pterophylla camellifolia* (common true katydid) is one of four species of true katydids that frequent oak-dominated forests.

but became less and less vociferous thereafter, falling completely silent from about 4 a.m. on. The overlapping "katy-did, katy-did" songs of many hundreds of these large orthopterans was a loud yet soothing white noise for me, something I looked forward to each summer, perhaps because its dependability help ground me during a time of rapid transition in my life.

Katydids belong to the family Tettigoniidae, the long-horned grasshoppers, a common name that refers to their extraordinarily long and delicate antennae. There are 262 species of tettigoniids in North America, but only four species of true katydids, the tettigoniids that spend their entire life in the canopy of oaks and other deciduous trees. It is the males in these species who are the singers, and they generate their songs through stridulation, a process in which one body part rubs against another. In the case of katydids, there is a scraper on the anterior end of one hindwing and a file at the same location on the other hindwing. By slightly lifting these wings and moving them rapidly back and forth, the scraper moves across the file, producing a surprisingly loud, species-specific song. Why so loud? Because long ago female katydids decided that loud males made better mates than quieter males. This was not a baseless evolutionary decision. Katydid song volume is correlated with male size; the loud males are large males, and that is what females are really after: large males that may have the best genes and certainly can pass females the most nutrients in their sperm packet (Gwynne 2001).

As in the vast majority of animal species, female katydids are fussy about their choice of mate. After all, they want the sire of their offspring to bear high-quality genes, and mating with just any old male leaves that goal to chance. But this creates a challenge for female katydids: how can one tell a male carrying high-quality genes from a male with inferior genes? Natural selection is good at solving these sorts of problems; rather than judging a male's gene quality directly, selection has favored females that demand a nuptial gift from hopeful suitors and then judge the quality of the gift. The assumption is that only high-quality males will be able to produce high-quality gifts. Anyone who expects a diamond ring on proposal night should be able to relate to this approach to mate choice. What would make a good gift for a female katydid? How about a package of food stuffed with protein? This works well for females in three ways: it is something that is easily assessed (large packages are better than small packages); package size accurately describes male quality (high-quality males are more likely to be able to make large packages than are low-quality males); and the nutrients in nuptial gifts can be shunted directly into developing eggs and are therefore useful for enhancing reproduction.

Here's how this all works. Male katydids sit in the canopy of your oak trees and sing. Females hear them with organs that function just like our own ears, that is, a hole with a membrane stretched tightly across its mouth that picks up vibrations in the air. There is a difference, however: katydid ears are in the tibia of the first pair of legs rather than the katydid's head. When katydids are singing, there usually are many males singing at once. This gives a receptive female a choice, and she invariably makes her way toward the male that is singing the loudest. When they meet, the female allows that male to attach a structure called a spermatophore to her genital opening.

After a male passes a spermatophore to a female, she immediately starts to eat part of it. But if it is large enough, the sperm it contains will have time to transfer to the female's genital tract before she eats it all.

The spermatophore is divided into two parts: an ampula, which is a small sac containing the male's sperm, and a spermatophylax, a much larger bag filled with nutrients. Two things happen as soon as the spermatophore is attached to the female. The ampula starts to pump sperm into the female, a process that takes about 20 minutes to complete. At the same time, the female bends her head under her body and starts to eat the spermatophylax. And now the race is on. If the spermatophylax is not big enough, the female will finish eating it before the ampula has emptied its sperm, and the female will eat the ampula and its contents as well. But if the male was able to make a large spermatophylax, the ampula will successfully transfer all its sperm to the female while she is still eating the spermatophylax. Thus, natural selection has favored, on the one hand, females that can identify the largest males with the best nuptial gifts (and presumably the best genes) and on the other hand, males that can grow a bit larger than their competitors and signal their size with a louder song. This form of sexual selection has served katydids well for millions of years and is shared by many of a katydid's fellow orthopterans.

The katydid species that sang me to sleep as a boy and the one most people are likely to encounter is *Pterophylla camellifolia*, the common true katydid. It is found throughout the treed regions of eastern North America, and even though it is a denizen of treetops, it often falls (true katydids cannot fly!) to the ground in the fall as it nears the end of its life. Females are easily distinguished from males by their slightly curved, flattened ovipositor, the structure protruding from the rear of the insect with which they glue large flat eggs inside of bark crevices. Because they spend their entire lives within sight of hungry birds, there has been intense selection for katydids that are colored like leaves, and some even have veination in their wings which looks just like leaf veins, midrib and all. This form of crypsis is especially astonishing in katydids in the tropics, where it is common to find species with wing coloration that mimics lichen patches, decayed leaf edges, caterpillar herbivory, or all three on the same wing. Mimicry in our North American species is modest in comparison but impressive nonetheless.

Slugs, saddlebacks, and other caterpillars

Many people have bucket lists—places and experiences they hope to go and have before they kick the bucket—and I am no exception. The items on my list, however, are more than a bit unorthodox. I want to see and photograph nature's most unusual or beautiful creations. This desire has made for a pretty long bucket list, but fortunately, many of these creatures occur right in our yard and I am slowly checking off my goals. One early member of my list was the caterpillar of the spun glass slug moth (*Isochaetes beutenmuelleri*). Not such an attractive name, to be sure, but few things in this world are more exquisitely beautiful, and I'll take an encounter with this moth's larva over a trip to the Taj Mahal any day.

Although I have found several on my oaks since, it was Brian Cutting, one of my graduate students at the time, who first introduced me to the spun glass slug. He was collecting data for his master's degree in our yard when he found a mature specimen of a spun glass caterpillar on one of our oaks. These caterpillars are so named because they have the bluish tint and lacy structure of artfully blown spun glass. Dozens of delicately arched filaments, each adorned with several glass-like balls, arise from the center line of the caterpillar and cascade over its sides like a fountain of blue water. *Isochaetes beutenmuelleri* belongs to the family Limacodidae, the slug cater-

pillars, whose larvae all somewhat resemble slugs in that their head is not visible from above, their legs are reduced in size, and they seem to glide slowly over leaves rather than walk, inch, or loop like other caterpillars. Those are their only similarities to true slugs, though, for they are not slimy in the least, and nearly all species are brilliantly colored.

When Brian found the spun glass slug he moved it into a jar so he could take it to me. I was thrilled and grabbed my camera for my first spun glass shots. I prefer to take nature photos in natural conditions, so I had to get the caterpillar out of the jar and onto an oak leaf. No problem, I thought; I'll just poke it a little

Larva and adult of the spun glass slug moth; the caterpillar looks, not surprisingly, like finely spun glass.

with my pencil to get it moving… poke, poke… Argh! Soon after my pencil touched one of its glassy filaments, the caterpillar dropped all the filaments from the rear half of its body, leaving a creature very unlike a spun glass slug! My photo op was ruined, but I had gained insight as to what the function of the filaments actually was. Protection against predators, most likely ants. If an ant attacked a spun glass slug, it would get nothing but a mouthful of glassy filament, and my guess is that when disturbed, the caterpillar also produces a foul smell, something a curious ant would rather not deal with. The caterpillar I had poked was fully grown, but if a younger caterpillar was forced to drop its shield of filaments, it would produce new ones during its next molt.

The spun glass slug is just one species of limacodid that I can readily find on our oaks. Beginning in July the early button slug, puss caterpillar, spiny oak slug, skiff moth, yellow-shouldered slug, hag moth, crowned slug, Nason's slug, smaller parasa, purple-crested slug, and saddleback caterpillar are all possibilities. In fact, there are 50 species of limacodids in North America, and the majority include oaks in their list of host plants. All are fascinating to watch, both as larvae and adults, but one word of caution: most limacodid caterpillars bear urticating hairs on their backs. Urticating hairs are more like spines that each have a small poison sac at their base. If you should bump up against one with enough force to break off a spine, the poison is released, and it produces a painful reaction much like that experienced from stinging nettles. In North America the saddleback caterpillar is by far our most frequently encountered limacodid, and of those I have brushed against, it is the most painful. But I wouldn't pet a puss caterpillar either, as I hear it is the most venomous caterpillar of them all!

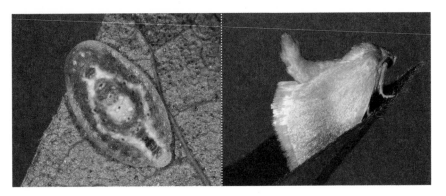

Early button, caterpillar and adult.

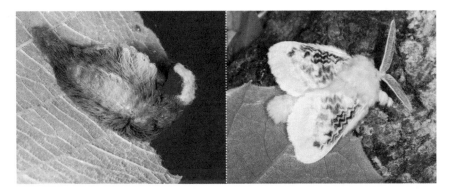

Puss caterpillar, southern flannel moth.

Spiny oak, caterpillar and adult.

Skiff, caterpillar and adult.

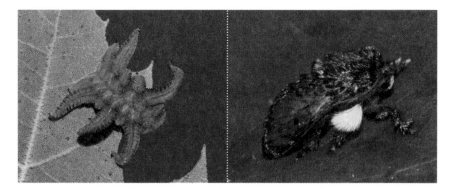

Hag, caterpillar and adult.

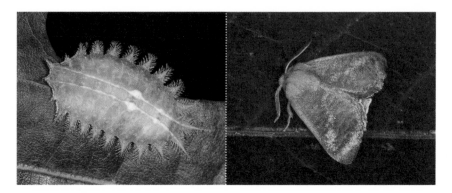

Crowned, caterpillar and adult.

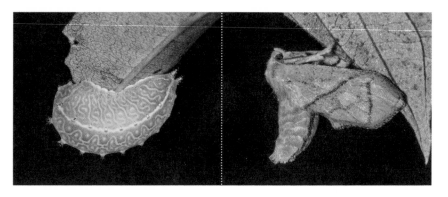

Nason's, caterpillar and adult.

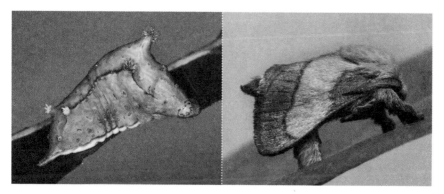

Smaller parasa, caterpillar and adult.

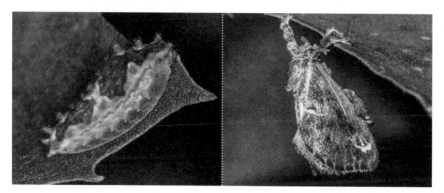

Purple-crested, caterpillar and adult.

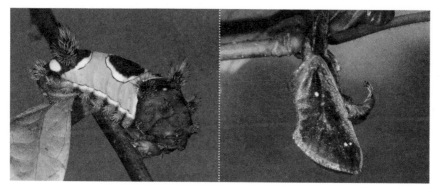

Saddleback, caterpillar and adult.

Gregarious feeders

By mid-July most birds have finished nesting, and they and their offspring start to supplement their diet of insects with berries and seeds. This diet switch relaxes the pressure on caterpillar populations, and they begin to increase in size as July proceeds. Probably the best time to look for caterpillars on your oaks east of the Mississippi is late July. In the West, caterpillar populations follow the rains, which have distinct cycles, particularly in the Southwest. In all areas except the upper Plains and Pacific Northwest, the most abundant caterpillar you are likely to encounter on oaks in July is the yellow-necked caterpillar (*Datana ministra*).

Like other datanas, adult females lay all their eggs at once and in one spot. Although tiny parasitic wasps immediately start laying their own eggs within the datana eggs, most of the datana eggs succeed in hatching, and the young caterpillars feed gregariously. Gregarious feeding is an adaptation to the increasing toughness of oak leaves as the season progresses, and it is a common feature of oak-feeding lepidopterans such as the spiny oakworm, the red-humped oakworm, the orange-humped oakworm, the pink-striped oakworm, the orange-striped oakworm, and the red-humped caterpillar, as well as *Datana* species, to name just a few. It is easier to bite through lignified oak leaves with 100 mouths than with one, so caterpillars forced to eat tough leaves tend to cram themselves onto a single leaf and eat as one. This type of feeding behavior concentrates caterpillar damage on one or two branches of your oak; your datanas may defoliate those branches, but as Tammany Baumgarten of New Orleans suggests, if you practice the 10-step program (that is, take 10 steps back from your oak), such feeding damage will disappear from sight!

Not every caterpillar species that uses your oaks as the leaves toughen up during July needs to feed gregariously like the yellow-necked caterpillar. The laugher (*Charadra deridens*), for example, demonstrates that there are always multiple ways for natural selection to solve a particular ecological problem. The laugher, so named because the black markings on its yellow "face" give it a permanent smile, has an unusually large head capsule. This is not because it is smarter than other caterpillars and has a bigger brain,

The yellow-necked caterpillar (*Datana ministra*) is a gregarious feeder on oaks. This helps the caterpillars overcome leaf toughness while they are small.

The laugher (*Charadra deridens*) overcomes oak leaf toughness with powerful mandibular muscles.

although, like many other oak-feeders, it is smart enough to hide from birds by day within a shelter it has built out of a leaf. Rather, the laugher has a big head to house large and powerful muscles that work its mandibles, enabling it to chew, solo, through even the toughest oak leaves. From Colorado and New Mexico east to the Atlantic, you can find the laugher by peeling open its leaf shelter. Once discovered, it will arch its back and give you a big smile.

For oaks, acorn size and shape matter

Near the end of July, fertilized ovules within the structures that were once oak flowers experience an explosive period of growth, so much so that they change from tiny nubbins along branches to what can clearly be recognized as an acorn. The cap (formally known as the cupule) develops first, and then the fleshy part of the acorn itself fills out over time beneath the cap. How energy is pumped into each acorn determines its eventual size and shape. And the size and shape of acorns is one of the most variable features of the genus *Quercus*.

Acorns don't vary in size and shape randomly; instead, they are shaped by many environmental and genetic factors. Let's start with acorn size. One thing that can influence acorn size is the dryness of the environment in which the oak species occurs; oaks that live in drier habitats, for example, tend to have small acorns, while those in wetter climates have enough moisture to grow larger acorns. Small acorns are also a feature of oaks in more northerly latitudes. It's possible that in the short growing seasons of the north there simply is not enough time for oaks to produce large acorns. Acorn size might also reflect the dispersal mode of the oak species; species with small acorns are more easily dispersed by jays and other birds. But small acorns are also a way of avoiding the ever-present acorn weevils; the smallest acorns are not big enough to provide enough nutrients for weevils to reach maturity. Finally, acorn size may be determined by the intensity of competition from other plants. Large acorns have enough stored energy to produce large radicles, the root-like structure that first emerges from a germinating acorn. Large radicles can quickly grow deep into the soil and tap resources beyond the reach of shallow-rooted competitors. This may

In the United States, acorns vary in size from the giants produced by bur oak (*Quercus macrocarpa*; top) to the smallest acorns made by Darlington oak (*Q. hemisphaerica*; bottom), with every size in between, represented here by northern red oak (*Q. rubra*).

explain the unusually large acorn of bur oaks found in oak savannahs of the Midwest, where competition from dense prairie plants might prevent the establishment of seedlings from small acorns.

Acorn shape does not vary as much as acorn size; for the most part, acorns are either round (spherical) or football-shaped (fusiform). The shape of acorns probably reflects a compromise between natural selection that favors a particular group of dispersers (mammals vs birds) and the ease with which an acorn can roll on the ground or be carried by water. Spherical acorns like those of many species in the red oak group are more easily picked up by squirrels, deer, chipmunks, and other mammals; fusiform acorns like those produced by live oaks (*Quercus virginiana*) are more easily handled by bird beaks and can more easily fit into bird crops. Suffice it to say that size and shape of the acorns produced by the oaks around you tell a story about the relationships your trees have with the animals and climate in your local environment.

There is one final way acorns vary across species, or at least across the broad taxonomic groups we call the red oaks and white oaks: they vary in the time it takes them to become mature seeds. Members of the white group all mature in one season (May–September), while red oak species require 18 months to mature. This difference in development time may explain why red oak species and white oak species rarely mast in the same year, serendipitously resulting in alternating years of surplus food for wildlife.

August

THROUGHOUT MOST OF the United States, summer is the season of thunderstorm deluges. In the East, strong thunderstorms that dump inches of rain in an hour or two can happen any time during the summer, but they are most common in July and August. The same is true for the Southwest's desert regions, where summer monsoons peak in late July and early August. Sudden, intense downpours can be destructive because water falls on the land faster than it can soak into the ground. Unimpeded heavy rain also pounds the ground hard enough to cause soil compaction. Anything that can break the force of heavy rain before it reaches the ground and/or absorb some of the water from a big storm can be considered a valuable ecosystem service. Oaks do both, and they do it really well.

Ecosystem services

So far we have considered some of the many biological interactions that accompany oaks wherever they exist, but I would be remiss if I didn't mention the abiotic benefits of adding oaks to your landscape, for these also have outsized effects on ecosystem health. Healthy ecosystems provide services that are vital, not only to our well-being but the well-being of all other plant and animal species in or near that ecosystem. Pollination; water purification; climate moderation; biodiversity maintenance; air pollution control; and the production of oxygen, food, fiber, and lumber are often cited as beneficial results derived from healthy ecosystems. Oaks contribute to several of these, but an underappreciated ecosystem service provided free of charge by oaks is watershed management.

When rain falls, it starts its inevitable journey to the sea. If this journey is rapid, the rain carries topsoil and pollutants with it to streams and then rivers, which terminate in the earth's oceans. If rainwater is slowed by vegetation, more of it seeps into the ground rather than rushing into local streams. Such infiltration not only replenishes water tables but also scrubs the water clean of its nitrogen, phosphorus, and heavy metal pollutants. Moreover, slow and steady discharge from water tables into streams and rivers reduces the destructive pulse of stormwater that scours our tributaries of their biota. By virtue of their copious leaf surface area and large root systems, oaks impede rainwater from the moment it condenses out of clouds. Much of the water intercepted by leafy oak canopies (up to 3,000 gallons per tree annually) evaporates before it ever reaches the ground (Cotrone 2014). All this makes oaks one of our very best tools in responsible watershed management.

Perhaps the most timely and critically important ecosystem service delivered every day by oaks is carbon sequestration. Like all plants, oaks fix atmospheric carbon dioxide (CO_2) through photosynthesis and store its carbon in their tissues. In fact, about half of a plant's dry weight (that is, its weight after all the water has been removed from its tissues) comes from carbon. For an average oak tree, this amounts to tons and tons of carbon. The more densely a plant's cells are packed together, the more carbon it can store; and it should come as no surprise that oaks produce some of the densest wood of all North American hardwoods. Oak contributions to below-ground carbon sequestration are also noteworthy. Like oak tissues above the ground, oak root systems are massive and built from carbon. But what makes oaks a particularly valuable tool in our fight against climate change is their relationship with mycorrhizal fungi: mycorrhizae make copious amounts of carbon-rich glomalin, a highly stable glycoprotein that gives soil much of its structure and dark color. Oak mycorrhizae deposit glomalin into the soil surrounding oak roots throughout the life of the tree. Every pound of glomalin produced by oak mycorrhizae is a pound of carbon no longer warming the atmosphere, and glomalin remains in soil for hundreds, if not thousands, of years. These factors rank oaks among our best options for scrubbing carbon from the atmosphere and storing it safely in soil throughout the world's temperate zones.

There is a common misconception promulgated by some people interested in using trees as a mechanism for removing CO_2 from the earth's atmosphere: fast-growing trees will remove more carbon than slow-growing trees (Korner 2017). In the short term, this is true, but if the goal is to remove CO_2 from the atmosphere long enough to reduce the greenhouse effect in any meaningful way, then the length of time the carbon is kept out of the atmosphere becomes a critical part of the equation. Removing carbon from the air quickly, only to release it back into the air a few decades later when the tree dies, just kicks the climate can a short way down the road to burden the next human generation. Thus, using fast-growing but short-lived trees like poplar and pine for carbon sequestration programs has no sustained effect on the amount of carbon in the atmosphere. Tree species that are ideal for carbon sequestration are large, long-lived, dense trees—like oaks—that will safely keep carbon out of the atmosphere for hundreds of years. Simply put, every oak you plant and nurture helps to moderate our rapidly deteriorating climate better than the overwhelming majority of plant species.

A final ecosystem service delivered by oaks that is worth considering is that oaks moderate our local microclimate in ways that make life more comfortable and energy-wise all year round. They block excessive wind, shade our houses in summer, and allow the sun to warm them in winter. Oak trees would also help reduce the heat island effect of cities during blistering heat waves if we would only plant more of them in urban areas.

The Bedford Oak in Bedford, New York, is over 500 years old, which is only midlife by some estimates.

Both the biotic and abiotic contributions that oak trees make to local ecosystems are a function of their age and size. Under ideal conditions of sunlight and water, many oak species can live more than a millennium as long as their roots are not obstructed by roads, sewer lines, house foundations, septic tanks, and so on. Some estimate that the Angel Oak in Charleston, South Carolina, for example, is over 1,500 years old. Healthy oaks will grow for 300 years, maintain a stasis between new growth and canopy loss for the next 300 years, and then decline for 300 years more. During each one of those 900 years, these magnificent plants are making outsized ecological contributions to the life around them. As oaks age they typically lose much of their inner xylem tissue, creating large hollow spaces within their trunks that serve as home to countless creatures, from rare fungi to raccoons, opossums, squirrels, bats, bobcats, and even black bears. We have been led to think that once there are hollow spaces created by rot within a tree trunk, that tree must come down. Not so! Such "rot" is normal and does not affect the living cambium that lies just under the bark of your oak nor the functional strength of the trunk. Hollow trunks are just one feature of ancient oaks that makes them such valuable ecological additions to our landscapes.

Beating oak defenses

By the time July slips into August, oak leaves have reached their toughest state and thus present their most formidable challenge to the insects that eat them. Large caterpillars with powerful mandibles are still able to eat oak leaves, although more slowly than in May, June, or July. But the lignin-filled outer layers of oak leaves are so tough by August that they successfully prevent most smaller caterpillar species from employing typical eating strategies. It seems ironic, then, that August is when the most species of small caterpillars do develop on oaks. Such species can do so because they have found ways to circumvent the armor in oak leaves. One way to do this is by avoiding the tough outer layers of oak leaves altogether.

If you were to cut an oak leaf in half, you would see that it looks a little like a sandwich. And like the bread layers of a sandwich, the upper and lower epidermis of the leaf protects the goodies between them—the soft and nutritious parenchymal cells of the palisade and spongy mesophyll. It is these innards of the oak leaf sandwich that are free for the taking to any creature that can reach them. Leaf miners have figured out how to do just that; by feeding only between the upper and lower epidermis, leaf miners can eat away at vulnerable leaf tissues just as if they were a seam of coal between two layers of sandstone. But, as with most things in life, there is a trade-off associated with the leaf mining mode of eating: you have to be really small to be a leaf miner!

An entire family of flies, the Agromyzidae, has specialized as leaf miners, but most species mine herbaceous plants; only a few agromyzids use oaks for development. One that does is the oak shothole leaf miner (*Japanagromyza viridula*). Larvae of this species form blotch mines that look just like their name: rounded blotches of hollowed-out areas within the oak leaves. You can distinguish oak shothole leaf miners from other types of blotch leaf miners by the presence of round holes in the leaves that line up in parallel on either side of a mine. Such holes are created by the adult flies rather than the larvae. When leaves are young, adult females pierce them with their ovipositors in order to drink the fluids that ooze out of the wound. As the leaves expand to their full size, the puncture wounds expand as well to become symmetrical holes in the leaves.

Although the oak shothole leaf miner is fairly common, most leaf miners on oaks are not flies but rather are tiny moths that create either blotch mines or serpentine mines. A serpentine mine is so called because it is thin and snakes its way across the leaf, expanding in width as the larva within it grows. As you might imagine, the caterpillars of moth leaf miners are tiny and so flattened they can eat their way between the upper and lower epidermis of oak leaves with no trouble at all. It follows that the adult moths that emerge from these mines are minute as well, but most are exquisitely colored nonetheless. Two closely related

The solitary oak leaf miner (*Cameraria hamadryadella*) is so tiny it can develop between the upper and lower layers of an oak leaf.

blotch miners exemplify this well: the solitary oak leaf miner (*Cameraria hamadryadella*) and the gregarious oak leaf miner (*C. cincinnatiella*) are both orange moths with three black stripes highlighted in white encircling the wings. As their names suggest, the solitary oak leaf miner makes only one blotch mine per leaf, while the gregarious oak leaf miner can have several blotch mines on a leaf. The mines are small, however, and not numerous enough to cause the tree to suffer from these moths in any meaningful way.

Slightly larger moths, like the stunning orange-headed epicallima (*Epicallima argenticinctella*), are able to chew through the upper leaf epidermis to reach the spongy mesophyll but do not bother with the lower epidermis. Even larger moths, though not large in comparison to most caterpillars we commonly encounter on oaks, fasten ropes of silk to one edge of a leaf and by pulling each one tight, manage to roll the leaf edge over to form a concealed dining room. Such leafrollers then skeletonize the leaf from within the protective confines of their rolled leaf. Species such as the white-lined leafroller (*Amorbia humerosana*) and the yellow-winged oak leafroller (*Argyrotaenia quercifoliana*) are just two of many common leafrollers on oaks. A final approach to leaf skeletonizing is employed by species such as the gold-striped leaftier (*Machimia tentoriferella*). Rather than rolling the edge of a single leaf as leafrollers do, leaftiers bind two overlapping oak leaves together with numerous silk strands and then, like the yellow-vested moth, skeletonize from within.

Enemies on the wing

For many invertebrate and vertebrate predators, particularly predatory insects, spiders, and birds, caterpillars are ideal sources of food and therefore ideal prey items. They are high in protein and fats, and the best source of essential carotenoids for creatures that don't eat plants. And because your oak tree harbors so many types of caterpillars, oaks are one of nature's most active stages for high drama as predators search oak nooks and crannies for caterpillars, and caterpillars do their very best to avoid becoming dinner. If caterpillar blood were red (it's green!), Alfred, Lord Tennyson's description of nature would be fully realized on oaks.

Though stink bugs, assassin bugs, damsel bugs, minute pirate bugs, and plant bugs are important predators of caterpillars and the eggs they hatched from, the most potent enemies of caterpillars all come from the sky. I speak of birds, parasitic wasps, solitary wasps, and social wasps. It is hard to approach an oak tree in August without seeing one or more birds poking about the leaves; potter wasps, yellowjackets, bald-faced hornets, and paper wasps meticulously searching leaf surfaces throughout the tree; and, if you have really good eyes, tiny parasitoids doing the same.

With all of this predation pressure, natural selection has woven defense strategies into every caterpillar's life history. Caterpillar shelters built from folded, rolled, or tied leaves; silk webs; or tiny stick houses are the most obvious ways caterpillars try to avoid being eaten, although none provide foolproof protection. This was brought home to me one day while I was counting caterpillars on the white oak in my front yard. I encountered a striped oak leaftier that seemed well defended to me. Not only had it created a tunnel-like shelter by tying three oak leaves together and always staying within, but it also had spun a dense web across any open space between leaves to block entrance by would-be predators or parasitoids. Yet, while I was watching it, a *Euodynerus leucomelas* potter wasp flew in, forced its

Potter wasps, yellowjackets, and bald-faced hornets hunt caterpillars on oaks from sunrise to sunset.

abdomen between silk strands, stung the caterpillar to immobilize it, and then chewed through enough of the web that it could drag the caterpillar, now paralyzed stiff as a board, out of its house. The wasp then flew off to its own shelter, carrying the caterpillar in its teeth (mandibles). Once home the wasp would lay a single egg on the living but helpless caterpillar, which would soon hatch into a legless, headless larva that would eat the caterpillar alive. As gruesome as this may seem, a paralyzed yet living caterpillar is the perfect way to preserve fresh food for your young for over a week without the luxury of a refrigerator.

Some of our most common caterpillars on oaks in August are members of the inchworm family, Geometridae. Inchworms are so called because they loop along while walking as if they were measuring out an inch with each step. Inchworms can mimic sticks both in form and function, and when disturbed, they will hold themselves at an angle out from a branch for minutes at a time, looking exactly like a small oak twig. As a teenager, I dabbled in gymnastic stunts and learned how much strength many gymnastic moves

 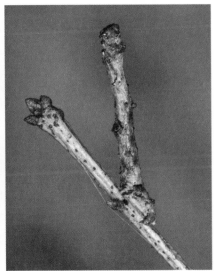

Inchworm stick mimics support their body weight with a silken tether, illuminated here by flash.

Oak beauty caterpillars are quintessential stick mimics.

required. One of the hardest moves (and one I never quite mastered) was to use my arms to hold my body out from a street sign post at a 45-degree angle. It demanded more arm and stomach strength than I had. And yet, that is exactly what stick-mimic inchworms seem to be doing. Whenever I saw one, I was in awe of how effortlessly it could arch its body away from the main branch, holding onto the branch with only its two hind prolegs. Or so I thought. The first time I photographed a stick mimic, the flash on my camera revealed the source of its herculean strength—a strand of silk tied to the tree branch. Stick mimics were not suspending their body with muscular strength at all, but rather just leaning back on a silken tether so thin it was invisible to the naked eye unless illuminated by a flash.

Looking like a stick is an effective way to hide from hungry birds, but it doesn't fool insect predators and parasitoids that hunt by smell rather than sight. Fortunately for many caterpillars, they have a plan B; if they are discovered by predators like ants, *Podisus* stink bugs, or cogwheel assassin bugs, they simply parachute off the leaf they were eating and hang

A parasitoid wasp searches for caterpillars on oak leaves day and night.

suspended in midair by a thread of silk. This seems to happen more at night than during the day, or at least is easier to see at night, and if you walk under your oak with a flashlight after dark, chances are good that you'll see several caterpillars hanging motionless a few inches below the nearest leaf. Each of these has dropped out of a predator's reach and will hang in the air until its enemy has left. Some species of parasitoid wasps, however, are not as easily discouraged; *Mesochorus discitergus*, for example, recognizes a silken thread as an opportunity and instead of giving up the hunt when their prey tumbles off the leaf, they use their front legs to pull the caterpillar up, hand over hand (or, more accurately, fore tarsi over fore tarsi), until it is close enough to lay an egg in (Yeargan and Braman 1989). Some parasitoids are even more nimble; while the caterpillar is hanging by its thread, these wasps shimmy down the thread itself and lay an egg in the caterpillar in midair!

Although mimicking twigs is an effective way to escape bird predation, it only works if you are a caterpillar that is skinny enough to look like a

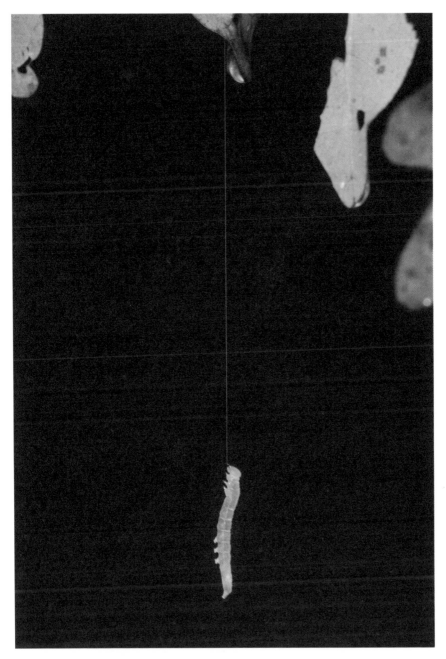

When threatened by predators or parasitoids, caterpillars often jump off their leaf and hang motionless from a silk thread until the danger has passed.

<small>CLOCKWISE FROM TOP LEFT</small> The checkered fringed prominent, lace-capped caterpillar, unicorn caterpillar, and red-washed prominent all resemble damaged oak leaves more than they do caterpillars.

twig. With so many birds hunting on oaks, caterpillars of all shapes (even the fat ones) have to protect themselves, and looking like leaf decay, leaf damage, bark, or the lichens that commonly grow on oaks is an excellent way to go about this. Caterpillars and their adult forms in the family Notodontidae are particularly good at mimicking oak tissues. The checkered fringed prominent (*Schizura ipomaeae*) is a wonderful example. Formerly

The adult moths of many notodontid species, like this red-washed prominent, are excellent twig mimics.

but mistakenly called the morning glory prominent (it does not eat morning glories), the checkered fringed prominent caterpillar can be found on leaf edges, looking remarkably like a section of leaf that has been munched by a caterpillar. Its body color includes the green of healthy leaf tissue as well as the mottled browns of dead leaf tissues, and its body shape mimics the wavy lines of caterpillar feeding damage. Similar body designs are found in other oak caterpillars, including the unicorn caterpillar (*S. unicornis*), the red-washed prominent (*Oligocentria semirufescens*), and the lace-capped caterpillar (*O. lignicolor*). The remarkable crypsis displayed by these species does not end at the caterpillar stage; as adults they mimic broken twigs in both looks and behavior.

Lace bugs and their predators

Tough leaves present an evolutionary challenge to insects that munch leaves, but many insect herbivores avoid such problems by sucking plant juices rather than chewing solid leaves. In so-called haustellate insects, the mandibles that allow caterpillars to chew through tough tissues have been modified into interlocking stylets that function like a straw. No matter how tough a leaf is, insects with sucking mouthparts can pierce the leaf epidermis and gain access to the nutritious goodies within the leaf. In insects such as aphids, the straw-like mouthparts can be quite long in relation to the size of the insect's head, but in others they are short and only able to suck juices from the cells near the surface of the leaf. This is the case in lace bugs, small heteropterans in the family Tingidae.

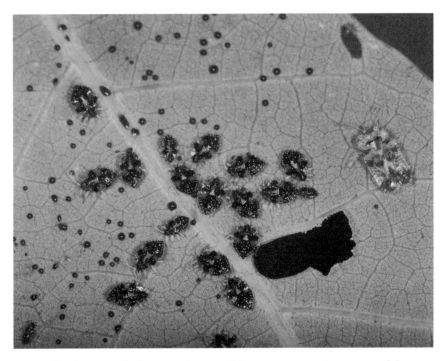

The oak lace bug (*Corythucha arcuata*) is one of the most common insects on oak leaves in August. Here an adult and many nymphs all share the same leaf.

Lace bugs are so named because their wings and prothorax appear to be constructed from a doily-like lace, which is quite beautiful when magnified. Most species are gregarious, and large groups of lace bug nymphs are often found mingling with adults and their egg masses, all on the same leaf. Several species of lace bugs specialize on oaks, but the most common in many parts of the country is the oak lace bug (*Corythucha arcuata*). Adults, which spend the winter in bark crevices, are one of the insects that nuthatches and brown creepers seek during the long winter months. Some lace bug adults always make it through the winter, though, and after oak leaves have fully expanded, they start the first of several generations by laying batches of 20 to 30 eggs on the underside of leaves. Oak lace bug populations build throughout the summer, and by mid-August, it can be hard to find a leaf that does not have one or more lace bugs on it. Though the adults are attractive under a microscope, nymphs deposit black fecal spots on the leaves

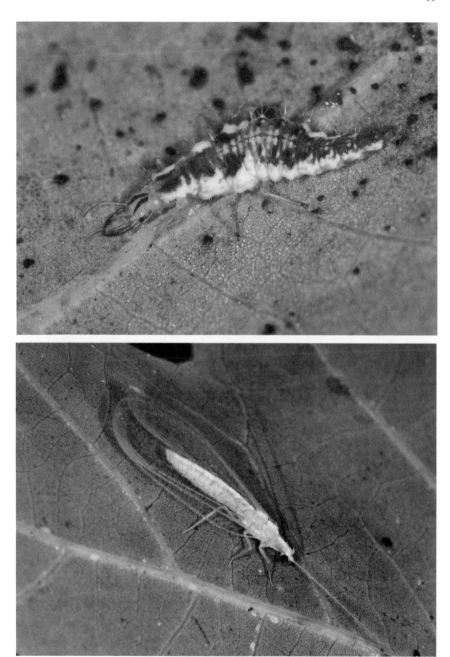

Lacewing larvae and adults are common predators on oaks when lace bug populations build.

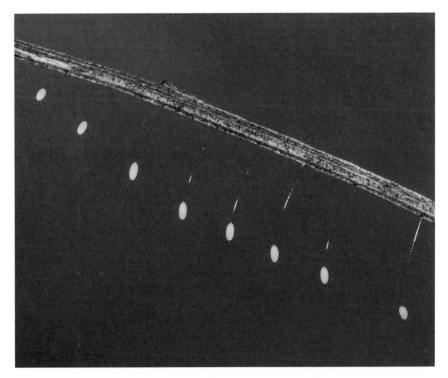

Each lacewing egg sits atop its own stalk, which prevents hatchlings from eating each other!

wherever they go, which, I'm sorry to say, does not enhance the aesthetics of the leaf. Large populations also remove a fair amount of chlorophyll from their host leaves, turning them from green to a light patchy tan. Please ignore this type of leaf damage. Oak lace bugs are an important part of the fauna using your oaks; they will not kill your tree, and treating them with insecticides will needlessly kill many nontarget species and add to global insect declines.

Large lace bug populations invariably attract lacewings (Chrysopidae), the principle predator of lace bugs. As adults, lacewings spend more time eating aphids than they do lace bugs, but lacewing larvae use large sickle-shaped mandibles to pierce and suck the innards of both lace bug eggs and nymphs—hundreds of lace bug eggs and nymphs throughout the life of each lacewing larva! Lacewing adults are attractive neuropterans, typically green and about an inch long. They hold lacy-veined wings roof-like over their abdomens and sport a delicate, benign appearance that belies the efficiency with which their larval stages dispatch immature lace bugs. Young lacewings are often called aphis lions because they also eat aphids whenever they encounter them. In fact, they eat just about any insect they can find, including their brothers and sisters. This has presented a challenge for mother lacewings: how to lay several eggs near a lace bug or aphid aggregation without encouraging offspring cannibalism? Natural selection was up to the challenge and provided a unique solution: elevate each egg on a tall stalk. When an egg hatches, the hungry lacewing larva drops to the leaf surface, where it is unable to reach its siblings for a quick snack.

Oak planthoppers

Other very common inhabitants of your oak trees that suck rather than chew their food are planthoppers in the families Flatidae and Acanaloniidae. As nymphs, both groups are globular in shape with big eyes and very short bodies. In fact, they appear to be heads with little legs. What distinguishes them from the immatures of other insects are dense strands of flocculent wax projecting from their rear ends in wavy strands longer than their entire body. This curious morphology is not without purpose. Over the eons, many insect lineages have found that, if they can make it, wax is excellent protection from predators. Some insects like alder woolly aphids, dogwood sawflies, and walnut woolly worms (another sawfly species) cover their entire body with a dense coating of distasteful wax. Others, like our oak planthoppers and insects in the family Fulgoridae produce most of their wax with glands at the terminal tip of their abdomen. These flimsy wax filaments are all that stand between predators approaching from the rear and the insect itself, but apparently a mouthful of wax is enough to deter ants, ladybird beetles, assassin bugs, and the many other types of planthopper enemies. Once they molt to their adult stage, planthoppers no longer protect themselves with flocculent wax, possibly because long waxy strands would interfere with flying, or because, as adults, they are excellent jumpers and can spring out of harm's way in a fraction of a second.

Adult and nymphal acanaloniid planthoppers can be very common on oaks in August.

Cicada killers

Depending on where you live, every 13 or 17 years June brings the emergence of thousands of periodical cicadas to yards with mature trees in the eastern United States. In contrast, regardless of what year it is, one or more species of annual cicadas begin to appear by mid-July and are on the wing through much of August. They are far less numerous than periodical cicadas, their background color of black is laced with green rather than the orange of periodical cicadas, and they are considerably larger than their periodical cousins. But, like all species of cicadas, annual cicada males court females with a loud buzzing song that is a central feature of lazy hot summer days throughout the United States.

Because annual cicadas are reliably present each summer, predators have been able to specialize on this dependable resource. Other than birds and squirrels that casually include cicadas in their diets, large wasps are

Large annual cicadas are the only prey of the cicada killer (*Sphecius speciosus*), our largest sphecid wasp.

the most unmistakable predator that eats only cicadas. These cicada killers (*Sphecius speciosus*) are among the scariest yet ironically most harmless wasps in North America. They certainly are the biggest members of the family Sphecidae and can approach 2 inches in length. Unlike social hornets and yellowjackets, sphecids like the cicada killer do not form hives containing a queen and hundreds of workers. Instead they are solitary wasps that cooperate with another wasp only while mating. There are times, though, when so many are seen flying in one area, people swear there is a hive nearby. What *is* nearby in these cases are the two resources cicada killers need to reproduce: diggable soil and a good population of annual cicadas.

A female cicada killer reproduces by capturing an annual cicada, a large insect itself, while in flight and stinging it into a paralyzed state. With a series of short flights and drags, it takes the cicada back to a vertical shaft she has previously dug in the ground. Cicada killer tunnels are substantial

A male cicada killer watches over his territory, ready to take flight to chase any intruders before they mate with "his" females.

underground excavations that require considerable effort on the part of the female to construct. They may be nearly a foot deep before they bend horizontally for another 6 inches, and they are wide enough to allow the wasp to drag a fat cicada down to a terminal chamber. Once the cicada is in place, the wasp lays an egg on it, then leaves the tunnel for good, sealing the entrance with dirt. The egg hatches in a few days, and the cicada killer larva consumes the cicada over the following weeks. Occasionally the female will provision her nest with more than one cicada for her young. When fully grown, the wasp larva pupates and spends the fall, winter, and spring waiting for the emergence of the first annual cicadas the following July before it ecloses to a fresh adult. Female cicada killers usually dig a new hole for each cicada they capture and will catch and bury cicadas throughout the four to five weeks of their adult life span (Alcock 1998).

Male cicada killers, like the males of most insects (indeed, most animals), contribute nothing to child care. They are interested in one thing: mating with as many female cicada killers as possible. Lucky for them that female cicada killers will mate after each nursery tunnel is stocked and sealed. And they will mate with the male that is strong enough to chase away all other males from the nesting site. This means males must constantly patrol areas where females are nesting—areas where the soil is just right for deep digging—so that they are on hand when a female emerges from a completed tunnel. And this is where we humans often misinterpret a male's intentions. Males, often to their detriment, are not very good at discriminating between other male cicada killers, which are legitimate rivals, and anything else that moves. They will chase your cat, your dog, your mailman, a dirt bomb tossed in their direction, or you to make sure you will not mate with the female of their dreams. If you are near an active nesting site, the males in that area take no chances; they fly straight toward you with a scowl on their harmless but admittedly scary faces. Remember, male wasps have no stinger, and they cannot hurt you in any way. But this fact is usually lost on homeowners whose yard happens to have the right type of soil for nesting wasps. Convinced they will be stung by these large wasps, too many homeowners pull out the can of Raid or pay an exterminator hundreds of dollars to kill all the wasps in their yard. Although female cicada killers do have a stinger, in my 45 years as a student of insect behavior I have never heard of any verified case of a cicada killer stinging a human. Females are focused on one thing: finding and burying the annual cicadas produced by your oak trees.

September

SEPTEMBER IS THE last month that the nutrition locked up in living oak leaves is available to leaf-eaters, and then just barely. By the time September rolls around, oak leaves are their thickest and toughest and contain the least amount of moisture. They also are rather ratty, having been fed upon by insects since they expanded in early May. Despite the declining nutritional value of oak leaves, September is still a good time to find and watch various creatures that use the oaks in your yard.

Sticks that walk

A group of insects that never fails to fascinate are the phasmid walking-sticks. Their name describes them well, for they are thin, linear insects that lack wings and look just like sticks. Phasmids, as a group, primarily hail from the tropics, where they are diverse and abundant and can achieve lengths exceeding a foot. In North America, however, there are only six species, and the largest is rarely over 5 inches in length. That is still an impressive length for an insect, and when they hold their front legs out in front of them, they are even longer.

Throughout the United States, the most commonly encounter phasmid is the northern walkingstick (*Diapheromera femorata*). It is a denizen of deciduous forests, where it skeletonizes the leaves of several tree species but shows a marked preference for white oak. Although most years its numbers remain low, in forests dominated by white oak, walkingsticks become numerous enough to cause noticeable herbivory every 10 years or so. Numerous or not, people rarely encounter phasmids for two reasons. First, their resemblance to sticks is so good that they are very difficult to distinguish from the millions of real sticks in a forest. Second, they spend most of their lives out of sight in the canopy of trees and are most active at night. In general, I see walkingsticks only during leaf fall in September and October, when they often climb up the side of my house after falling out of my oaks.

Phasmids have several lines of defense against vertebrate predators like birds and squirrels. The most obvious is crypsis; simply blending in with their stick-filled habitat works well under most circumstances. But there is an important behavioral component to their crypsis that makes it even more effective. Leaves and thin sticks on trees are not rigid and motionless; they constantly move in ways that we find entirely natural with breezes or the sudden jarring from a predator landing on a nearby branch. If a walkingstick were to remain stiff when it should be shaking like the rest of its branch, it might be noticed for what it really is—a tasty insect. Instead, when it detects an approaching predator, a walkingstick quivers and quakes as if it were a plant part. If this fails to convince a predator that it is an inedible stick, the insect drops to the forest floor, where it becomes stick-rigid and motionless among the thousands of real sticks. For walkingsticks unlucky enough to be handled by a predator, there is one final line of defense. Several species have glands that release a chemical noxious to most predators. A few species, like the two-striped walkingstick (*Anisomorpha buprestoides*), can actually shoot their chemical defense several inches from glands on their prothorax, often straight into the eyes of an unsuspecting bird!

An Arizona walkingstick (*Diapheromera arizonensis*) picks its way slowly through the foliage of an Emory oak.

Walkingstick females mature in late summer; they then mate and begin to lay eggs. Each large, seed-like egg is unceremoniously dropped by its mother as she moves through the forest canopy to the leaf litter below. In this way, eggs are scattered on the ground throughout the forest, where they remain during the fall, winter, and early spring. In late spring walkingstick nymphs hatch from many (but not all) of the eggs, and the hatchlings eat leaves they encounter as they climb laboriously to the tops of the nearest trees. The eggs that don't hatch remain on the forest floor another entire year and hatch the following spring. This reproductive strategy is a perfect example of ecological bet hedging; by staggering egg hatch over the course of two years, walkingsticks reduce the chances that all their young will be lost to unpredictable adverse conditions like severe drought, hurricane, fire, floods, or even rare but extreme cataclysms like the asteroid that destroyed the dinosaur empire 66 million years ago.

Duskywings

Other creatures whose second generation can be fairly common on oak leaves in September include one or more species of duskywing skippers. Unlike most skippers, Juvenal's duskywing (*Erynnis juvenalis*), Horace's duskywing (*E. horatius*), sleepy duskywing (*E. brizo*), Zarucco duskywing (*E. zarucco*), and Propertius duskywing (*E. propertius*) are all mottled brown as adults, and telling one species from another in the field is nigh on impossible. Their caterpillars, however, can be quite attractive and uniquely marked. Skippers are an evolutionary enigma that share some traits of butterflies (diurnal flight, recurved antennae) and moths (stout body). The majority of skipper species are specialized to eat grasses, and their larval morphology reflects adaptations to deal with the toughness of typical grass foliage. Most grass species have copious amounts of silica in their tissues and can only be consumed by insects endowed with powerful mandibles and the musculature to operate them. In skipper caterpillars, these muscles are housed within the head capsule. Unusually large heads are therefore standard equipment for this group of lepidopterans; but, curiously, the region just behind a skipper's head—its neck, if you will—is exceptionally narrow as caterpillar necks go, making skippers look very much like they have just been strangled!

For reasons we can only guess, oak duskywings have abandoned the grass-eating behavior of their ancestors and now specialize on oaks. The adaptations they accumulated over time for eating tough grass leaves serve them well as oak specialists, for as we have seen, oak leaves are as tough as old grass stems throughout most of the season. Like all other skippers, oak duskywings build leaf shelters by folding over a section of leaf and tying it in place with silk. These serve as safe sites by day and occasionally places to dine in peace. A single caterpillar will build and abandon several leaf shelters as it grows, and these can be used to track down these interesting insects. When they have completed their development, the caterpillars drop to the ground and pupate within the leaf litter at the base of your oak.

This brings up an important point about the lepidopterans that use oak leaves for growth and reproduction. Only a few species complete their entire

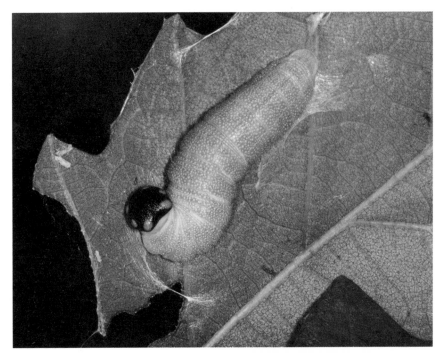

Juvenal's duskywing (*Erynnis juvenalis*) hides by day within a folded leaf but ventures forth at night to eat oak foliage.

life cycle on their oak host itself; over 90% of the hundreds of caterpillar species that use oaks drop from their host tree when they have completed their larval development and either pupate underground or spin a cocoon in the leaf litter under their oak tree. Unfortunately, our typical landscaping habits conflict with both of these necessary activities. Most of us do not allow leaf litter to accumulate under our oaks and, instead, mow grass over compacted soil right up to the trunks of our trees. This practice doesn't help our trees or the community of soil organisms under our trees, and it's absolutely deadly for most oak caterpillars. When we don't allow caterpillars to complete their life cycle, we create what is called an ecological trap; the oak draws in female moths to lay their eggs, only to have their surviving larvae chopped up by our lawn mowers, squished under our feet, or die trying to work their way underground within soil too dry and compacted to permit digging.

Beds of plants under your oaks rather than lawn are far better for your tree and also will allow the caterpillars on your tree to complete their development.

An easy solution to this dilemma is to create beds under our trees that remain undisturbed all year long. This is an ideal place to establish a bed of spring ephemerals or to display beautiful ground covers like wild ginger, ferns, Virginia creeper, various species of violets, or native pachysandra. Better yet, build a three-dimensional landscape by planting understory trees like hazelnuts, dogwoods, witchhazels, or ironwoods under your oak, possibly underplanted by shrubs like native azaleas, viburnums, and blueberries. Oak leaf litter under these plantings would complete a perfect habitat for oak caterpillars.

Tree crickets

The various cricket species in our yards have all reached maturity by late summer, and so the night chorus of their mating songs reaches its peak on warm September evenings. Most cricket species inhabit the ground, low herbaceous vegetation, or shrubs, but a group called tree crickets typically sing from the leaves of trees. One you can readily find on your oaks across the country is the snowy tree cricket (*Oecanthus fultoni*). Typically a pale green, the snowy tree cricket can be so pale it appears white (hence "snowy").

Snowy tree crickets have been called temperature crickets because the air temperature can be estimated with surprising accuracy by counting the rate at which they chirp. Although he didn't mention the species of cricket he worked with, later studies suggest it was the snowy tree cricket that prompted Amos Dolbear to conduct a crude experiment in 1897. While listening to crickets near his home, Dolbear noted that at 60 degrees Fahrenheit, snowy tree crickets made 80 chirps per minute, at 70 degrees they chirped 120 times per minute, and at 50 degrees, they only chirped 40 times per minute. These measurements allowed Dolbear to establish the following relationship: temperature F = 40 + N, where N is the number of chirps per 15 seconds (Dolbear 1897). This simple equation has come to be known as Dolbear's law, and with slight modifications it can be applied to other species of crickets as well. Even if you can't see the crickets chirping in your oak tree this fall, you can use the weather app and timing device on your smartphone to see just how accurate Dolbear's observations were more than a century ago.

Snowy tree crickets may be able to tell us the temperature, but other tree cricket species that inhabit our oaks have equally impressive tricks. Males of the two-spotted tree cricket (*Neoxabea bipunctata*), for example, have discovered over evolutionary time that if they sit at the edge of a hole in a leaf with their head just inside the hole, and then lift their wings over their heads so that their wings nearly fill the hole, the sound produced when they sing, a continuous high-pitched whir, is amplified. Not just any old hole will do; it must approximate the area of both wings combined. When this condition is met, the leaf hole acts like an acoustic baffle and allows males

Tree cricket species like *Neoxabea bipunctata* (at rest, top) use a hole in a leaf to project a louder, clearer signal while singing (bottom).

to produce a sound much louder than their body size would permit if they were to sing without aid of the leaf hole. Remember, females judge male quality by body size, which they assess indirectly by the volume of the songs males produce. Tree crickets in Central America have gone one step further than the two-spotted tree cricket; rather than spending time searching for a leaf hole just the right size, they chew a hole of the correct specifications in a leaf that faces the direction they wish to amplify their song. You may find falsely advertising your actual size to receptive females a bit deceptive, but it reminds me of the 20-year-old guy who has no money in the bank yet buys the fanciest car he can on credit to impress his dates with his apparent wealth. It's not just cricket males that are willing to stretch the truth a bit to catch the eye of an attractive female.

Preparing for winter

Though September is still warm in most areas of the country, the official first day of fall is 21 September, and signs of the coming winter are unmistakable. Goldenrods have started to transition from flower to seed; echinaceas, evening primroses, and black-eyed Susans are in full seed set, as are most other plants that flowered during the summer. In fact, seeds of all kinds are never more abundant than they are in September. This is a true bounty for seed-eaters but one that is synchronized, ephemeral, and available only once during the year. And so, with frenetic efficiency, granivores of all types—including squirrels, chipmunks, mice, and many bird species— must gather as much of this bonanza as possible and "squirrel" it away for later use during the cold winter months before the seed stocks are gone.

You can watch this drama unfold if you put your bird feeder up in September. For a foraging bird, nothing is more convenient than a well-stocked feeder, and they will favor that source of food over everything except a juicy insect. Watch, for example, the chickadees and titmice at your feeder. They do not sit and eat seed after seed the way many finches do. Instead, they dart in, grab a seed, and fly off, usually out of sight. But what then? What are your chickadees and titmice doing with each of those seeds? They are hiding them in secret places only they know about. Over time, these birds build up caches of seeds in reliable safe sites, which they will visit throughout the winter for food. Seed-cachers don't know that you will be stocking your feeder all winter long (except when you go to Florida for a week in January). They treat your feeder just like any other temporary seed supply they might find in September: grab as many seeds as possible before something else takes them, and hide them somewhere safe for later use. And this is where your oak tree comes in handy. Many oak species, but particularly those in the white oak group, have slightly shaggy bark with thousands of nooks and crannies perfect for slipping seeds into. Older oaks are also a rich source of cavities created by lost branches or foraging woodpeckers. All such hiding places are potential safe sites for birds to cache their fall seed supplies.

As with people, some birds are sneakier than others and are not above taking advantage of the hard work of seed-cachers. Blue jays (and other jay species) are the most notorious in this regard. They will sit quietly and watch the comings and goings of chickadees and titmice, and then go and eat the "hidden" seeds almost as fast as those tiny birds store them. All the more reason to create landscapes that provide enough seeds for workers and cheaters alike.

Epilogue

IN THE PRECEDING CHAPTERS I have described some of the life that is associated with oaks in general, and the oak trees I have planted in our yard in particular, over the course of a year. None of the animals I have mentioned would be able to thrive in our yard had I not planted those oaks, and neither would any of the species that, in turn, depend on those oak-associated animals. My survey of the species that absolutely require the oaks in our yard or just include them as one of their resources has hardly been exhaustive; to discuss all the thousands of known animal interactions with oaks would have quickly exceeded my attention span as an author and, I'm betting, yours as a reader. Instead, my intention was for this book to serve as a starting point, perhaps pique your interest in oak communities, or at least increase your appreciation of the central role oaks play in ecosystems across the country, and indeed, ecosystems in the northern hemisphere the world over.

If you are at all interested in contributing to the conservation of local animals, or in enjoying the wonders of nature right at home, planting one or more oaks is an awfully good way to do those things.

Although oaks can live to become ancient cornerstones of ecosystems throughout the United States, the old giants that once provided so many unique niches for layers upon layers of biodiversity are now largely absent from our landscapes. Wherever oaks occurred, they were prized suppliers of wood products, and most large specimens were logged centuries ago. Long after the giants were gone, we continued to degrade oak habitats across the country. Vast tracts of oak forests have been "developed" (from an ecological perspective, this is one of our most ironic words), converted to crop or pastureland, or have been highly altered by fire suppression. In fact, the percentage of oaks in eastern forests has dropped from 55% pre–European settlement to 25% today (Hanberry and Nowacki 2016). Add pressure from the collapse of the climate that favored oak health over the last 8,000 years, as well as the human introduction of diseases like sudden oak death, oak wilt, and oak leaf scorch, and invasive pests like the gypsy moth, and many oak species are now on the ropes. A recent analysis by the Morton Arboretum in Lisle, Illinois, has found that 28 of the 91 oak species in North America (over 30%) are so diminished in numbers, they may soon disappear from the wild forever (Morton Arboretum 2015). The noble Oregon white oak (*Quercus garryana*), for example, has lost 97% of its habitat in the last 200 years.

We cannot casually accept the loss of oaks without also accepting the loss of thousands of other plants and animals that depend on them. Oak declines in the United Kingdom, for example, threaten the survival of some 2,300 other species (Mitchell et al. 2019). Fortunately, there is no reason why we *should* accept the loss of oaks as inevitable; there is no trick to restoring oak populations, and no shortage of places in which to restore them. If you were to add up the amount of land in various types of built landscapes that is not dedicated to agriculture—suburban developments, urban parks, golf courses, mine reclamation sites, and so forth—it would total 603 million acres, a full 33% of our lower 48 states. We have not targeted these places for conservation in the past, but that was back when our conservation model was based on the notion that humans and their tailings were here and nature was someplace else. That model of mutual exclusion has failed us dismally; there simply are not enough untrammeled places left to sustain the natural world that until now has sustained us. Our only option, then, is to find ways to coexist with other species. That's right, we must construct ecosystems that contain all their functional parts right where humans abound.

The Oregon white oak (*Quercus garryana*) is just one of many species whose populations have been reduced by over 90%.

This may turn out to be easier than you think. From the perspective of effective conservation, it is fortuitous that many of our human-dominated landscapes structurally resemble the highly diverse, sun-drenched ecosystems that once dominated North America (think suburbia!). That squirrel we have heard about that could run through treetops from the Atlantic Ocean to the Mississippi River without ever touching the ground may never have existed. There is growing evidence that deciduous forests throughout the United States (in fact, temperate zones the world over) were historically much more open than the dark, closed canopy forests of today, resembling savannahs dotted with a diversity of large spreading trees as well as a rich assemblage of sun-loving prairie plants and all the insects, birds, mammals, reptiles, and amphibians that depended on them (Mitchell 2004). What kept these forests open, it is thought, were the enormous and abundant Pleistocene mammals that dominated North America for millions of

Many present-day open spaces resemble the savannah-like structure of North American landscapes before the Pleistocene mammals were hunted to extinction.

years: mammals like the American mastodon that stood 10 feet high at the shoulder; giant ground sloths that could reach vegetation 15 feet high and more; herds of giant bison, 4,000-pound beasts that dwarfed extant bison; camels, horses, tapirs, peccaries—the list goes on and on. The elimination of these browsers at the end of the Pleistocene allowed forests in the northern hemisphere to become more dense and shaded than they ever were in the past, suppressing prairie plant species that require lots of light. Where once there were hundreds of plant species per acre, there are now relatively few. My point is, we may be able to restore much of the biodiversity of old without restructuring the physical aspects of where we live, work, and play now, simply by adding the right plants to our existing landscapes.

We humans live our lives out in a brief instant of ecological time. We cannot return ancient oaks to our landscapes during that instant, but we can—indeed, we must—start the process. I have planted a number of massive old oaks on our property, except they are only 19 years old and are not so massive yet. They are growing, though, and several have topped 30 feet at this writing. In a blink of time they will be large enough and old enough to fully assume their keystone positions in our yard. And just in time. Our winner-take-all extraction-based economies, coupled with our unchecked population expansion, have triggered the sixth great extinction event of planet earth's 4-billion-year history. Over a million species are headed for extinction in next few years unless we take immediate action (Sartore 2019). We have no choice but to prevent such losses, not because we're nice guys, but because it is those species that run the ecosystems that support us. And by "we," I mean every single human earth-dweller, not just the surprisingly few people who already recognize the necessity of sustainable earth stewardship. Although their impending demise is bad enough, it is not the loss of rare and endangered species that we should fear; their populations are no longer large enough to have major impacts on our ecosystems. Rather, it is the loss of the common kingpins like oaks that we must prevent as if our well-being depends on them. For it does.

Acknowledgments

THERE WERE SECTIONS of this book that may seem relatively unimportant to the reader but which required heroic efforts above and beyond the call of friendship to complete. The section on Temnothorax ants, which make nests within acorns after acorn weevils have left, stands out. I needed a photo of the ants themselves. I'm not much of an ant guy, but my friend Steve Vail is. I casually mentioned that if he ran across any Temnothorax colonies, could he please save them for me. Not only did he make it his business to find the desired ants, but he also flew them from his home in New Jersey to a small airport near me in his private plane! Same-day delivery, better than Amazon!

A second noble contribution to this book was made by renowned oak expert Guy Sternberg. I made a late decision to add a section about acorn variability, but I made that decision well after acorns had dropped across the country. I asked Guy if he happened to have any acorns for pictures. He didn't, but he knew folks who did. In the end he personally collected acorns from Missouri, Georgia, and Louisiana and took the photo I needed himself. I am beyond grateful to both Steve and Guy.

Other difficult-to-obtain photos were supplied by macrophotographers extraordinaire Sara Bright (great purple hairstreaks) and Dave Funk (katydids). I am honored to know them both.

I am also indebted to Kimberley Shropshire, my steadfast research assistant, who helped with every research project reported in this book, and to my (now former) graduate students Desiree Narango, Ashley Kennedy, and Adam Mitchell, who tolerated my lack of attention whenever I became distracted by my writing, which was often. Finally, I would not have written one word of this or any other book had it not been for encouragement and essential facilitation from my wife, Cindy. What a world this would be if everyone were so lucky!

References

Alcock, J. 1998. "Taking the sting out of wasps." *American Gardener* 77: 20–21.

Angst, Š. T., et al. 2017. "Retention of dead standing plant biomass (marcescence) increases subsequent litter decomposition in the soil organic layer." *Plant and Soil* 418:571–579.

Bailey, R. K., et al. 2009. "Host niches and defensive extended phenotypes structure parasitoid wasp communities." *PLoS Biology* 7:1–12.

Bossema, I. 1979. "Jays and oaks: An eco-ethological study of symbiosis." *Behaviour* 70:1–117.

Condon, M. A., et al. 2008. "Hidden neotropical diversity: greater than the sum of its parts." *Science* 320:928–931.

Cotrone, V. 2014. "A green solution to stormwater management." Penn State Extension. extension.psu.edu/a-green-solution-to-stormwater-management.

Dirzo, R., et al. 2014. "Defaunation in the Anthropocene." *Science* 345:401–406.

Dolbear, A. E. 1897. "The cricket as a thermometer." *The American Naturalist* 31:970–971.

Faaborg, J. 2002. *Saving Migrant Birds*. Austin: University of Texas Press.

Forister, M. L., et al. 2015. "Global distribution of diet breadth in insect herbivores." *Proceedings of the National Academy of Sciences* 112:442–447.

Grandez-Rios, J. M., et al. 2015. "The effect of host-plant phylogenetic isolation on species richness, composition and specialization of insect herbivores: a comparison between native and exotic hosts." *PLoS ONE* 10:e0138031.

Griffith, R. 2014. "Marcescence." Youtube.com (video with narration).

Gwynne, D. T. 2001. *Katydids and Bush-crickets*. Ithaca, N.Y.: Comstock Publishing Associates.

Hanberry, B. B., and G. J. Nowacki. 2016. "Oaks were the historical foundation genus of east-central United States." *Quaternary Science Reviews* 145:94–103.

Heinrich, B., and R. Bell. 1995. "Winter food of a small insectivorous bird, the Golden-crowned Kinglet." *Wilson Bulletin* 107:558–561.

Janzen, D. H. 1968. "Host plants as islands in evolutionary and contemporary time." *The American Naturalist* 102:592–595.

———. 1973. "Host plants as islands, ii: competition in evolutionary and contemporary time." *The American Naturalist* 107:786–790.

Kelly, D., and V. L. Sork. 2002. "Mast seeding in perennial plants: why, how, where?" *Annual Review of Ecology and Systematics* 33:427–447.

Kerlinger, P. 2009. *How Birds Migrate*. Mechanicsburg, Pa.: Stackpole Books.

Koenig, W., and J. Knops. 2005. "The mystery of masting in trees." *American Scientist* 93:340.

Koenig, W. D., et al. 2018. "Effects of mistletoe (*Phoradendron villosum*) on California oaks." *Biology Letters* 14:20180240.

Korner, C. 2017. "A matter of tree longevity." *Science* 355:130–131.

Krauss, J., and W. Funke. 1999. "Extraordinary high density of Protura in a windfall area of young spruce plants." *Pedobiologia* 43:44–46.

Logan, W. B. 2005. *Oak: The Frame of Civilization*. New York: W. W. Norton.

Mitchell, F. J. G. 2004. "How open were European primeval forests? Hypothesis testing using paleoecological data." *Journal of Ecology* 93:168–177.

Mitchell, R. J., et al. 2019. "Collapsing foundations: the ecology of the British oak, implications of its decline and mitigation options." *Biological Conservation* 233:316–327.

Morton Arboretum. 2015. mortonarb.org/science-conservation/global-tree-conservation/projects/global-oak-conservation-partnership.

Narango, D., et al. 2017. "Native plants improve breeding and foraging habitat for an insectivorous bird." *Biological Conservation* 213:42–50.

———. 2018. "Nonnative plants reduce population growth of an insectivorous bird." *Proceedings of the National Academy of Sciences* 115:11549–11554.

Ostfeld, R. S., et al. 1996. "Of mice and mast." *BioScience* 46:323–330.

Pearse, I. S., et al. 2016. "Mechanisms of mast seeding: resources, weather, cues, and selection." *New Phytologist* 212: 546–562.

Platt, H. M. 1994. Foreword to *The Phylogenetic Systematics of Free-living Nematodes* by S. Lorenzen and S. A. Lorenzen. London: The Ray Society.

Ponge, J., et al. 1997. "Soil fauna and site assessment in beech stands of the Belgian Ardennes." *Canadian Journal of Forest Research* 27:2053–2064.

Richard, M., D. W. Tallamy, and A. Mitchell. 2018. "Introduced plants reduce species interactions." *Biological Invasions* 21:983–992.

Rosenberg, K. V., et al. 2019. "Decline of North American avifauna." *Science* 366:120–124.

Sartore, J. 2019. "One million species at risk of extinction, UN report warns." nationalgeographic.com/environment/2019/05/ipbes-un-biodiversity-report-warns-one-million-species-at-risk/#close.

Sourakov, A. 2013. "Two heads are better than one: false head allows *Calycopis cecrops* (Lycaenidae) to escape predation by a jumping spider, *Phidippus pulcherrimus* (Salticidae)." *Journal of Natural History* 47:15–16.

Southwood, T. R. E., and C. E. J. Kennedy. 1983. "Trees as islands." *Oikos* 41:359–371.

Svendsen, C. R. 2001. "Effects of marcescent leaves on winter browsing by large herbivores in northern temperate deciduous forests." *Alces* 37:475–482.

Sweeney, B. W., and J. D. Newbold. 2014. "Streamside forest buffer width needed to protect stream water quality, habitat, and organisms." *Journal of the American Water Resources Association* 50:560–584.

Sweeney, B. W., and J. G. Blaine. 2016. "River conservation, restoration, and preservation: rewarding private behavior to enhance the commons." *Freshwater Science* 35:755–763.

Tallamy, D. W. 1999. "Child care among the insects." *Scientific American* 280:50–55.

Tallamy, D. W., and W. P. Brown. 1999. "Semelparity and the evolution of maternal care in insects." *Animal Behavior* 57:727–730.

Yeargan, K. V., and S. K. Braman. 1989. "Life history of the hyperparasitoid *Mesochorus discitergus* and tactics used to overcome the defensive behavior of the green cloverworm." *Annals of the Entomological Society of America* 82:393–398.

How to plant an oak

WHETHER YOU ADD OAKS to your landscape by planting acorns, small bareroots, or larger balled-and-burlapped trees, knowing a few basics about planting oaks can make the difference between failure and success. Regardless of which route you take, step one is choosing an appropriate site for your oak. The most important thing to remember is that oaks grow, and contrary to urban legend, many species grow quite quickly. With this in mind, be sure to allocate enough space for your oak so that it can seamlessly join the other plants in your landscape when it is a mature tree. It takes a little imagination to picture your landscape 10, 20, 30 years down the road, but it's worth this planning exercise. Those years fly by faster than any of us think they will.

Planting time is also when you should consider adding a grove of oaks rather than isolated singletons. As I described in the February chapter, two or three trees planted within 10 feet of each other will interlock their roots as they grow and therefore be far less vulnerable to dangerous blowdowns when they mature. These are not unnaturally tight spacings in our forests—just in our minds.

Although I may not succeed, I am going to try to convince you to plant acorns rather than transplant established trees. Acorns are easy, free, and plentiful, and they will grow into healthier trees than if you transplant established trees. The only thing they don't give you is instant gratification. But if you do decide to try acorns, here are a few hints to help you out. First, know which species you are planting. Members of the white oak group germinate in the fall just days after dropping from the parent tree. These should be planted soon after collecting, about a half-inch deep (pretend you are a squirrel or a blue jay storing your nut for the winter). Your acorns will send down their taproot during the fall but will not appear aboveground, in the form of an epicotyl, until spring.

In contrast, members of the red oak group do not germinate until the following spring. Even so, don't wait until spring to collect your red oak acorns; usually the only ones you will find in the spring will be inferior or damaged acorns already rejected by myriad acorn-eaters over the wintertime. As with acorns from the white oak group, collect your red oak acorns soon after they drop in the fall and either plant them right away, or store them in a sealed plastic bag filled with damp peat moss in the refrigerator until mid-March. This will protect them from hungry rodents during the long winter months.

This brings us to an important decision when planting acorns. If you plant them directly into the ground at their finally resting site, chances are good they will be found and destroyed by some critter before they germinate. For this reason, I plant my acorns in deep flowerpots in a mixture of your local soil and potting mix that drains well. You will then need to protect your pot from marauding mice, preferably someplace that is exposed to some of winter's cold but not the full might of a polar vortex. Remember, your pot won't receive any moderating warmth from the unfrozen soil beneath it, as it would if your acorn were planted directly in the ground.

Another winter challenge is desiccation; if you store your pot where it will not be exposed to seasonal rain or snow, your acorn may dry out. You can avoid this problem by lightly watering about once a month. In the spring, wait until your seedling has fully expanded its first true leaves before transplanting into the ground. But don't wait too long. Oaks grown in pots quickly become rootbound, a condition that often kills the tree months or years after transplanting. This all may sound pretty complicated, but it's really not that hard and, in my view, beats spending the thousands of dollars that would otherwise go toward buying a large tree.

Speaking of large trees, this is the most common way people acquire oaks, but it comes with serious downsides. For the sake of instant gratification, and because of the mistaken notion that if you plant a small tree, you'll be dead before you get to enjoy it, many people are willing to spend many thousands of dollars to buy a 3- to 4-inch caliper oak. Again, there are so many disadvantages to this approach that installing large oaks is at the very bottom of my "how to get an oak" list. But it is an option.

There is a reasonable compromise between planting an acorn and buying a large tree and that is to buy an inexpensive bareroot whip. Bareroot whips are heavily pruned, dormant trees just a few feet tall. All the dirt has been removed from their roots (hence, "bareroot"), so they are light and easily transported and only cost a few dollars each. You can store your bareroot for a few days before planting as long as you keep the roots moist within a plastic bag. It's best to plant bareroots in early spring so they can break dormancy in sync with the season. Dig a hole one-third wider than the roots as they would spread naturally from the trunk, but be careful not to dig the hole deeper than the crown, the area on the trunk where the first roots appear. Planting a tree below its crown is the most common way to kill it. Some people like to dig a deep hole and then backfill it to the proper depth, but this should be avoided as well. Backfilled soil will settle over time, resulting in a tree whose crown is below grade. Once you are sure the tree is not too deep, fill your hole while holding the tree upright in place. Jiggle the roots up and down a bit to make sure the soil fills in all the spaces among the roots, and water your planting liberally the first day or two.

Depending on where you live, your young oak will probably need protection from mammals like rabbits and deer. I use a 5-foot-high circular wire cage of galvanized steel that is wide enough so the young tree can spread its branches without being bent out of shape by the cage. If you are starting from an acorn, you'll want to increase the cage size as the small tree grows. I admit these cages are unsightly, but until we get our deer populations under control, I have found them to be the most foolproof way of preventing deer damage. Few things are more infuriating than to nurse along an oak for several years only to have a deer snip off its leader because you took the cage off too soon. Your oak will outgrow dangers from deer soon enough, at which point you can celebrate graduation day—the day you remove the cage for good!

Best oak options for your area

Northeast = ME NH VT MA CT RI

LARGE TREES

Q. alba (northern white oak): ME NH VT MA CT RI
Q. macrocarpa (bur oak): ME NH VT; rare in CT, western MA
Q. montana (chestnut oak): MA CT RI, southern NH; rare in western VT, southern ME
Q. palustris (pin oak): MA CT RI
Q. rubra (northern red oak): ME NH VT MA CT RI
Q. stellata (post oak): CT, eastern MA; rare in RI
Q. velutina (black oak): ME NH VT MA CT RI

INTERMEDIATE-SIZED TREES

Q. bicolor (swamp white oak): MA CT RI, western VT; rare in ME
Q. coccinea (scarlet oak): MA CT RI; rare in VT, southern ME
Q. muehlenbergii (chinkapin oak): western CT; rare in VT, western MA

SMALL TREES

Q. ilicifolia (bear oak): NH MA CT RI, southern ME; rare in southern VT
Q. prinoides (dwarf chinkapin oak): VT MA CT RI, southern NH

Mid-Atlantic = NY PA NJ DE MD VA

LARGE TREES

Q. alba (northern white oak): NY PA NJ DE MD VA

Q. falcata (southern red oak): DE MD VA, southern NJ; rare in southeastern PA

Q. hemisphaerica (Darlington oak): rare in southeastern VA

Q. imbricaria (shingle oak): PA MD; rare in southern NJ, western VA

Q. laevis (turkey oak): rare in southeastern VA

Q. laurifolia (laurel oak): southeastern VA

Q. macrocarpa (bur oak): NY PA; rare in MD, western VA

Q. montana (chestnut oak): NY PA NJ DE MD VA

Q. nigra (water oak): DE MD, eastern VA; rare in NJ

Q. pagoda (cherry-bark oak): southern MD, eastern VA; rare in DE

Q. palustris (pin oak): NY PA NJ DE MD VA

Q. phellos (willow oak): NJ DE MD VA; rare in PA, Long Island NY

Q. rubra (northern red oak): NY PA NJ DE MD VA

Q. shumardii (Shumard's oak): VA; rare in MD PA, western NY

Q. stellata (post oak): NY PA NJ DE MD VA

Q. velutina (black oak): NY PA NJ DE MD VA

Q. virginiana (live oak): rare in southeastern VA

INTERMEDIATE-SIZED TREES

Q. bicolor (swamp white oak): NY PA NJ DE MD VA

Q. coccinea (scarlet oak): NY PA NJ DE MD VA

Q. lyrata (overcup oak): southern MD, eastern VA; rare in DE, southern NJ

Q. marilandica (blackjack oak): MD VA, southeastern PA; rare in DE, Long Island NY

Q. michauxii (swamp chestnut oak): DE MD VA; rare in NJ, southeastern PA

Q. muehlenbergii (chinkapin oak): NY PA MD, northern DE; rare in NJ

SMALL TREES

Q. ilicifolia (bear oak): NY PA NJ MD, western VA

Q. incana (bluejack oak): southern VA

Q. margarettae (sand post oak): rare in southeastern VA

Q. prinoides (dwarf chinkapin oak): NY PA; rare in DE MD VA

Southeast = NC SC GA

LARGE TREES

Q. alba (northern white oak): NC SC GA

Q. arkansana (Arkansas oak): rare in GA

Q. falcata (southern red oak): NC SC GA

Q. hemisphaerica (Darlington oak): SC, eastern NC, southern GA

Q. imbricaria (shingle oak): rare in western NC, northern GA

Q. laevis (turkey oak): eastern NC, southern SC GA

Q. laurifolia (laurel oak): NC SC GA

Q. montana (chestnut oak): western NC SC, northern GA

Q. nigra (water oak): NC SC GA

Q. oglethorpensis (Oglethorpe oak): rare in western SC, northern GA

Q. pagoda (cherry-bark oak): NC SC GA

Q. palustris (pin oak): SC; rare in NC

Q. phellos (willow oak): NC SC GA

Q. rubra (northern red oak): NC SC, northern GA

Q. shumardii (Shumard's oak): NC SC GA

Q. similis (bottomland post oak): rare in eastern SC GA

Q. sinuata (bastard oak): rare in SC GA

Q. stellata (post oak): NC SC GA

Q. velutina (black oak): NC SC GA

Q. virginiana (live oak): SC, eastern NC, southern GA

INTERMEDIATE-SIZED TREES

Q. austrina (bluff oak): rare in NC SC GA

Q. bicolor (swamp white oak): rare in NC SC

Q. coccinea (scarlet oak): NC SC GA

Q. geminata (sand live oak): coastal plains NC SC GA

Q. georgiana (Georgia oak): central GA; rare in NC SC

Q. lyrata (overcup oak): NC SC GA

Q. marilandica (blackjack oak): NC SC GA

Q. michauxii (swamp chestnut oak): NC SC GA

Q. muehlenbergii (chinkapin oak): SC GA; rare in NC

Q. myrtifolia (myrtle oak): southeastern GA; rare in SC

SMALL TREES

Q. chapmanii (Chapman's oak): rare in SC, eastern GA

Q. ilicifolia (bear oak): rare in western NC

Q. incana (bluejack oak): NC SC GA

Q. inopina (sandhill oak): rare in SC, eastern GA

Q. margarettae (sand post oak): NC SC GA

Q. minima (dwarf live oak): eastern SC, southern GA; rare in
southeastern NC

Q. prinoides (dwarf chinkapin oak): western SC; rare in western NC,
northern GA

Q. pumila (running oak): SC GA; rare in southeastern NC

Gulf Coast Region = FL AL MS LA

LARGE TREES

Q. alba (northern white oak): FL AL MS LA

Q. arkansana (Arkansas oak): rare in AL MS LA, western FL

Q. falcata (southern red oak): FL AL MS LA

Q. hemisphaerica (Darlington oak): FL AL MS LA

Q. imbricaria (shingle oak): northern AL; rare in LA, northern MS

Q. laevis (turkey oak): FL, southern AL, southeastern MS; rare in eastern LA

Q. macrocarpa (bur oak): rare in AL MS LA

Q. montana (chestnut oak): AL, northern MS

Q. nigra (water oak): FL AL MS LA

Q. oglethorpensis (Oglethorpe oak): rare in AL MS LA

Q. pagoda (cherry-bark oak): AL MS LA, western FL

Q. palustris (pin oak): AL; rare in northern MS

Q. phellos (willow oak): AL MS LA, northern FL

Q. rubra (northern red oak): AL MS; rare in LA

Q. shumardii (Shumard's oak): AL MS LA, northern FL

Q. similis (bottomland post oak): MS LA; rare in southern AL

Q. sinuata (bastard oak): AL MS, northern FL; rare in LA

Q. stellata (post oak): AL MS LA, northern FL

Q. texana (Texas red oak): AL MS LA

Q. velutina (black oak): AL MS LA, northern FL

Q. virginiana (live oak): FL AL MS LA

INTERMEDIATE-SIZED TREES

Q. austrina (bluff oak): AL MS, northern FL

Q. bicolor (swamp white oak): rare in northern AL

Q. coccinea (scarlet oak): AL MS; rare in eastern LA

Q. geminata (sand live oak): FL, southern AL MS LA

Q. georgiana (Georgia oak): rare in AL

Q. lyrata (overcup oak): AL MS LA, northern FL

Q. marilandica (blackjack oak): AL MS LA, western FL

Q. michauxii (swamp chestnut oak): AL MS LA, northern FL

Q. muehlenbergii (chinkapin oak): AL MS LA, western FL

Q. myrtifolia (myrtle oak): FL, southern AL; rare in southern MS

SMALL TREES

Q. boyntonii (Boynton's sand post oak): rare in central AL

Q. chapmanii (Chapman's oak): FL, southern AL

Q. incana (bluejack oak): FL AL MS LA
Q. inopina (sandhill oak): FL, southern AL
Q. margarettae (sand post oak): FL AL MS LA
Q. minima (dwarf live oak): FL; rare in southern AL MS
Q. prinoides (dwarf chinkapin oak): AL MS LA
Q. pumila (running oak): FL, southern AL MS

Southwest = TX OK NM AZ

LARGE TREES

Q. alba (northern white oak): eastern TX OK
Q. arkansana (Arkansas oak): rare in eastern TX
Q. falcata (southern red oak): eastern TX OK
Q. hemisphaerica (Darlington oak): eastern TX
Q. imbricaria (shingle oak): easternmost TX
Q. laurifolia (laurel oak): eastern TX
Q. macrocarpa (bur oak): TX OK; rare in northwestern NM
Q. nigra (water oak): eastern TX OK
Q. pagoda (cherry-bark oak): eastern TX OK
Q. palustris (pin oak): eastern OK
Q. phellos (willow oak): eastern TX OK
Q. polymorpha (net-leaf white oak): rare in southeastern TX
Q. rubra (northern red oak): eastern OK
Q. shumardii (Shumard's oak): eastern TX OK
Q. similis (bottomland post oak): eastern TX
Q. sinuata (bastard oak): TX; rare in southern OK
Q. stellata (post oak): TX OK
Q. texana (Texas red oak): eastern TX; rare in eastern OK
Q. velutina (black oak): eastern TX OK
Q. viminea (Sonoran oak): rare in southeastern AZ
Q. virginiana (live oak): TX; rare in southern OK

INTERMEDIATE-SIZED TREES

Q. buckleyi (Buckley's oak): TX OK

Q. carmenensis (Mexican oak): rare in western TX

Q. chrysolepis (canyon live oak): AZ; rare in southwestern NM

Q. gambelii (Gambel's oak): NM AZ, western TX; rare in western OK

Q. gravesii (Graves' oak): western TX

Q. grisea (gray oak): NM AZ, western TX

Q. lyrata (overcup oak): eastern TX OK

Q. marilandica (blackjack oak): OK, eastern TX

Q. michauxii (swamp chestnut oak): eastern TX OK

Q. muehlenbergii (chinkapin oak): TX OK, eastern NM

Q. oblongifolia (Mexican blue oak): NM, southern AZ

Q. pungens (sandpaper oak): NM, southern AZ, western TX

Q. robusta (robust oak): rare in western TX

Q. rugosa (net-leaf oak): western TX, southwestern NM, southern AZ

Q. vaseyana (Vasey's oak): central & western TX

SMALL TREES

Q. ajoensis (Ajo Mountain oak): southern AZ

Q. arizonica (Arizona white oak): NM AZ, western TX

Q. canbyi (Chisos oak): rare in western TX

Q. chihuahuensis (Chihuahuan oak): rare in western TX

Q. depressipes (Davis Mountain oak): rare in western TX

Q. emoryi (Emory oak): NM AZ, western TX

Q. fusiformis (plateau oak): TX, southern OK

Q. havardii (Havard's oak): northwestern TX, western OK, eastern NM

Q. hinckleyi (Hinckley's oak): western TX

Q. hypoleucoides (silver-leaf oak): southeastern AZ, southwestern NM; rare in western TX

Q. incana (bluejack oak): eastern TX; rare in eastern OK

Q. intricata (dwarf oak): western TX

Q. laceyi (Lacey's oak): central & western TX

Q. margarettae (sand post oak): OK, eastern TX

Q. mohriana (Mohr's oak): western TX OK, eastern NM

Q. palmeri (Dunn's oak): AZ, southwestern NM

Q. prinoides (dwarf chinkapin oak): OK

Q. tardifolia (late-leaf oak): rare in western TX

Q. toumeyi (Toumey's oak): southwestern NM, southern AZ; rare in western TX

Q. turbinella (shrub live oak): NM AZ; rare in western TX

Q. welshii (shinnery oak): northwestern NM, northern AZ

Midwest = WV KY TN MO AR OH IN IL IA

LARGE TREES

Q. acerifolia (maple-leaf oak): rare in western AR

Q. alba (northern white oak): WV KY TN MO AR OH IN IL IA

Q. arkansana (Arkansas oak): rare in southwestern AR

Q. ellipsoidalis (northern pin oak): IA, northern OH IN IL

Q. falcata (southern red oak): WV KY TN, southern MO IN IL; rare in southern OH

Q. hemisphaerica (Darlington oak): southernmost AR

Q. imbricaria (shingle oak): WV KY TN MO AR OH IN IL, southern IA

Q. laurifolia (laurel oak): AR, south-central TN

Q. macrocarpa (bur oak): WV KY MO AR OH IN IL IA, western TN

Q. montana (chestnut oak): WV KY TN OH, southern IN; rare in southern IL, southeastern MO

Q. nigra (water oak): TN; rare in KY, southeastern MO

Q. pagoda (cherry-bark oak): AR TN, southeastern MO, western KY, southern IL; rare in southern IN

Q. palustris (pin oak): WV KY TN MO AR OH IN IL, southern IA

Q. phellos (willow oak): KY TN AR, southeastern MO; rare in southern IL

Q. rubra (northern red oak): WV KY TN MO AR OH IN IL IA

Q. shumardii (Shumard's oak): KY TN MO AR OH IN IL; rare in WV

Q. sinuata (bastard oak): rare in southern AR

Q. stellata (post oak): WV KY TN MO AR, southern OH IN IL IA

Q. texana (Texas red oak): AR, western TN; rare in western KY, southern IL, southeastern MO

Q. velutina (black oak): WV KY TN MO AR OH IN IL IA

Q. virginiana (live oak): rare in southwestern TN

INTERMEDIATE-SIZED TREES

Q. austrina (bluff oak): rare in southwestern AR

Q. bicolor (swamp white oak): WV KY TN MO IN IL IA

Q. coccinea (scarlet oak): WV KY TN MO OH IN IL, eastern AR

Q. lyrata (overcup oak): KY TN AR, southeastern MO, southern IL; rare in southern IN

Q. marilandica (blackjack oak): WV KY TN MO AR IL, southern IN IA; rare in southern OH

Q. michauxii (swamp chestnut oak): KY TN AR, southeastern MO, southern IL IN

Q. muehlenbergii (chinkapin oak): WV KY TN MO AR OH IN IL IA

SMALL TREES

Q. ilicifolia (bear oak): eastern WV

Q. incana (bluejack oak): southeastern AR

Q. prinoides (dwarf chinkapin oak): TN MO, southern IA; rare in KY IN

Plains States = KS NE SD ND

LARGE TREES

Q. alba (northern white oak): eastern KS

Q. ellipsoidalis (northern pin oak): easternmost ND

Q. imbricaria (shingle oak): easternmost KS

Q. macrocarpa (bur oak): KS NE SD ND

Q. palustris (pin oak): eastern KS

Q. rubra (northern red oak): eastern KS NE

Q. shumardii (Shumard's oak): eastern KS; rare in southeastern NE

Q. stellata (post oak): eastern KS, southeastern NE

Q. velutina (black oak): eastern KS, southeastern NE

INTERMEDIATE-SIZED TREES

Q. bicolor (swamp white oak): rare in eastern NE

Q. gambelii (Gambel's oak): rare in NE

Q. marilandica (blackjack oak): eastern KS; rare in southeastern NE

Q. muehlenbergii (chinkapin oak): eastern KS, southeastern NE

SMALL TREES

Q. havardii (Havard's oak): rare in southern KS

Q. prinoides (dwarf chinkapin oak): eastern KS; rare in southeastern NE

Upper Midwest= MI WI MN

LARGE TREES

Q. alba (northern white oak): MI WI MN

Q. ellipsoidalis (northern pin oak): MI WI MN

Q. imbricaria (shingle oak): southern MI

Q. macrocarpa (bur oak): MI WI MN

Q. montana (chestnut oak): southeastern MI

Q. palustris (pin oak): southern MI; rare in WI

Q. rubra (northern red oak): MI WI MN

Q. shumardii (Shumard's oak): rare in southern MI

Q. velutina (black oak): MI WI, eastern MN

INTERMEDIATE-SIZED TREES

Q. coccinea (scarlet oak): southern WI

Q. muehlenbergii (chinkapin oak): southern MI; rare in southern WI

SMALL TREES
Q. prinoides (dwarf chinkapin oak): southern MI

Rocky Mountain Region = MT ID WY CO UT

LARGE TREES
Q. macrocarpa (bur oak): northeastern WY; rare in southeastern MT

INTERMEDIATE-SIZED TREES
Q. gambelii (Gambel's oak): CO UT, southern WY; rare in southern ID
Q. grisea (gray oak): southern CO

SMALL TREES
Q. turbinella (shrub live oak): CO UT
Q. welshii (shinnery oak): western CO, southeastern UT

Southwest California = CA NV

LARGE TREES
Q. agrifolia (coastal live oak): CA
Q. lobata (valley oak): CA
Q. tomentella (island live oak): rare in southwestern CA
Q. wislizeni (interior live oak): CA

INTERMEDIATE-SIZED TREES
Q. chrysolepis (canyon live oak): CA, western NV
Q. douglasii (blue oak): CA
Q. gambelii (Gambel's oak): southern NV
Q. garryana (Oregon white oak): CA
Q. kelloggii (California black oak): CA

SMALL TREES

Q. berberidifolia (California scrub oak): CA

Q. cedrosensis (Cedros Island oak): rare in southern CA

Q. cornelius-mulleri (scrub oak): southern CA

Q. dumosa (coastal sage scrub oak): CA

Q. durata (leather oak): CA

Q. engelmannii (Engelmann's oak): rare in southern CA

Q. john-tuckeri (Tucker's oak): southern CA

Q. pacifica (Pacific oak): rare in southwestern CA

Q. palmeri (Dunn's oak): central & southern CA

Q. parvula (Santa Cruz Island oak): coastal CA

Q. turbinella (shrub live oak): southern NV; rare in southwestern CA

Q. vacciniifolia (huckleberry oak): CA, western NV

Pacific Northwest= WA OR & northern CA

LARGE TREES

Q. agrifolia (coastal live oak): CA

Q. lobata (valley oak): CA

Q. wislizeni (interior live oak): CA

INTERMEDIATE-SIZED TREES

Q. chrysolepis (canyon live oak): CA, southwestern OR

Q. douglasii (blue oak): CA

Q. garryana (Oregon white oak): CA, western OR WA

Q. kelloggii (California black oak): CA, southwestern OR

SMALL TREES

Q. berberidifolia (California scrub oak): CA

Q. dumosa (coastal sage scrub oak): CA

Q. durata (leather oak): CA

Q. parvula (Santa Cruz Island oak): coastal CA

Q. sadleriana (deer oak): CA, southwestern OR

Q. vacciniifolia (huckleberry oak): CA, southwestern OR

North American native oaks

Q. acerifolia (maple-leaf oak): rare in western AR

Q. agrifolia (coastal live oak): CA

Q. ajoensis (Ajo Mountain oak): southern AZ

Q. alba (northern white oak): ME NH VT MA CT RI NY PA NJ DE MD
VA NC SC GA FL AL MS LA MI WI MN WV KY TN MO AR OH
IN IL IA, eastern TX OK KS

Q. arizonica (Arizona white oak): NM AZ, western TX

Q. arkansana (Arkansas oak): rare in GA AL MS LA, western FL, eastern
TX, southwestern AR

Q. austrina (bluff oak): AL MS, northern FL; rare in NC SC GA,
southwestern AR

Q. berberidifolia (California scrub oak): CA

Q. bicolor (swamp white oak): MA CT RI NY PA NJ DE MD VA WV KY
TN MO IN IL IA, western VT; rare in ME NC SC, northern AL,
eastern NE

Q. boyntonii (Boynton's sand post oak): rare in central AL

Q. buckleyi (Buckley's oak): TX OK

Q. canbyi (Chisos oak): rare in western TX

Q. carmenensis (Mexican oak): rare in western TX

Q. cedrosensis (Cedros Island oak): rare in southern CA

Q. chapmanii (Chapman's oak): FL, southern AL; rare in SC, eastern GA

Q. chihuahuensis (Chihuahuan oak): rare in western TX

Q. chrysolepis (canyon live oak): AZ CA, western NV, southwestern OR; rare in southwestern NM

Q. coccinea (scarlet oak): MA CT RI NY PA NJ DE MD VA NC SC GA AL MS WV KY TN MO OH IN IL, eastern AR, southern WI; rare in VT, southern ME, eastern LA

Q. cornelius-mulleri (scrub oak): southern CA

Q. depressipes (Davis Mountain oak): rare in western TX

Q. douglasii (blue oak): CA

Q. dumosa (coastal sage scrub oak): CA

Q. durata (leather oak): CA

Q. ellipsoidalis (northern pin oak): IA MI WI MN, northern OH IN IL, easternmost ND

Q. emoryi (Emory oak): NM AZ, western TX

Q. engelmannii (Engelmann's oak): rare in southern CA

Q. falcata (southern red oak): DE MD VA NC SC GA FL AL MS LA WV KY TN, southern NJ MO IN IL, eastern TX OK; rare in southeastern PA, southern OH

Q. fusiformis (plateau oak): TX, southern OK

Q. gambelii (Gambel's oak): NM AZ CO UT, southern WY NV, western TX; rare in NE, southern ID, western OK

Q. garryana (Oregon white oak): CA, western OR WA

Q. geminata (sand live oak): FL, coastal plains NC SC GA, southern AL MS LA

Q. georgiana (Georgia oak): central GA; rare in NC SC AL

Q. gravesii (Graves' oak): western TX

Q. grisea (gray oak): NM AZ, western TX, southern CO

Q. havardii (Havard's oak): northwestern TX, western OK, eastern NM; rare in southern KS

Q. hemisphaerica (Darlington oak): SC FL AL MS LA, eastern NC TX, southern GA, southernmost AR; rare in southeastern VA

Q. hinckleyi (Hinckley's oak): western TX

Q. hypoleucoides (silver-leaf oak): southeastern AZ, southwestern NM; rare in western TX

Q. ilicifolia (bear oak): NH MA CT RI NY PA NJ MD, southern ME, western VA, eastern WV; rare in southern VT, western NC

Q. imbricaria (shingle oak): PA MD WV KY TN MO AR OH IN IL, southern IA MI, easternmost TX KS, northern AL; rare in LA, western VA NC, northern GA MS, southern NJ

Q. incana (bluejack oak): NC SC GA FL AL MS LA, southern VA, southeastern AR, eastern TX; rare in eastern OK

Q. inopina (sandhill oak): FL, southern AL; rare in SC, eastern GA

Q. intricata (dwarf oak): western TX

Q. john-tuckeri (John-Tucker's oak): southern CA

Q. kelloggii (California black oak): CA, southwestern OR

Q. laceyi (Lacey's oak): central & western TX

Q. laevis (turkey oak): FL, southern AL SC GA, southeastern MS, eastern NC; rare in southeastern VA, eastern LA

Q. laurifolia (laurel oak): NC SC GA AR, southeastern VA, south-central TN, eastern TX

Q. lobata (valley oak): CA

Q. lyrata (overcup oak): NC SC GA AL MS LA KY TN AR, eastern VA TX OK, southern MD IL, southeastern MO, northern FL; rare in DE, southern NJ IN

Q. macrocarpa (bur oak): ME NH VT NY PA TX OK WV KY MO AR OH IN IL IA KS NE SD ND MI WI MN, western TN, northeastern WY; rare in CT MD AL MS LA, western VA MA, northwestern NM, southeastern MT

Q. margarettae (sand post oak): NC SC GA FL AL MS LA OK, eastern TX; rare in southeastern VA

Q. marilandica (blackjack oak): MD VA NC SC GA AL MS LA OK WV KY TN MO AR IL, southern IN IA, eastern TX KS, western FL, southeastern PA; rare in DE, Long Island NY, southern OH, southeastern NE

Q. michauxii (swamp chestnut oak): DE MD VA NC SC GA AL MS LA KY TN AR, southern IL IN, eastern TX OK, southeastern MO, northern FL; rare in NJ, southeastern PA

Q. minima (dwarf live oak): FL, eastern SC, southern GA; rare in southeastern NC, southern AL MS

Q. mohriana (Mohr's oak): western TX OK, eastern NM

Q. montana (chestnut oak): MA CT RI NY PA NJ DE MD VA AL WV KY TN OH, southern NH IN, western NC SC, northern GA MS, southeastern MI; rare in southern ME IL, western VT, southeastern MO

Q. muehlenbergii (chinkapin oak): NY PA MD SC GA AL MS LA TX OK WV KY TN MO AR OH IN IL IA, western FL CT, eastern KS NM, northern DE, southeastern NE, southern MI; rare in VT NJ NC, southern WI, western MA

Q. myrtifolia (myrtle oak): FL, southern AL, southeastern GA; rare in SC, southern MS

Q. nigra (water oak): DE MD NC SC GA FL AL MS LA TN, eastern VA TX OK; rare in NJ KY, southeastern MO

Q. oblongifolia (Mexican blue oak): NM, southern AZ

Q. oglethorpensis (Oglethorpe oak): rare in AL MS LA, western SC, northern GA

Q. pacifica (Pacific oak): rare in southwestern CA

Q. pagoda (cherry-bark oak): NC SC GA AL MS LA AR TN, eastern VA TX OK, western FL KY, southern MD IL, southeastern MO; rare in DE, southern IN

Q. palmeri (Dunn's oak): AZ, southwestern NM, central & southern CA

Q. palustris (pin oak): MA CT RI NY PA NJ DE MD VA SC AL WV KY TN MO AR OH IN IL, southern IA MI, eastern OK KS; rare in NC WI, northern MS

Q. parvula (Santa Cruz Island oak): coastal CA

Q. phellos (willow oak): NJ DE MD VA NC SC GA AL MS LA KY TN AR, eastern TX OK, northern FL, southeastern MO; rare in PA, Long Island NY, southern IL

Q. polymorpha (net-leaf white oak): rare in southeastern TX

Q. prinoides (dwarf chinkapin oak): VT MA CT RI NY PA AL MS LA OK TN MO, southern NH IA MI, western SC, eastern KS; rare in DE MD VA KY IN, western NC, northern GA, southeastern NE

Q. pumila (running oak): SC GA FL, southern AL MS; rare in southeastern NC

Q. pungens (sandpaper oak): NM, southern AZ, western TX

Q. robusta (robust oak): rare in western TX

Q. rubra (northern red oak): ME NH VT MA CT RI NY PA NJ DE MD VA NC SC AL MS WV KY TN MO AR OH IN IL IA MI WI MN, eastern OK KS NE, northern GA; rare in LA

Q. rugosa (net-leaf oak): western TX, southwestern NM, southern AZ

Q. sadleriana (deer oak): northern CA, southwestern OR

Q. shumardii (Shumard's oak): VA NC SC GA AL MS LA KY TN MO AR OH IN IL, eastern TX OK KS, northern FL; rare in MD PA WV, western NY, southern MI, southeastern NE

Q. similis (bottomland post oak): MS LA, eastern TX; rare in eastern SC GA, southern AL

Q. sinuata (bastard oak): AL MS TX, northern FL; rare in SC GA LA, southern AR OK

Q. stellata (post oak): CT NY PA NJ DE MD VA NC SC GA AL MS LA TX OK WV KY TN MO AR, southern OH IN IL IA, eastern MA KS, northern FL, southeastern NE; rare in RI

Q. tardifolia (late-leaf oak): rare in western TX

Q. texana (Texas red oak): AL MS LA AR, eastern TX, western TN; rare in western KY, southern IL, southeastern MO, eastern OK

Q. tomentella (island live oak): rare in southwestern CA

Q. toumeyi (Toumey's oak): southwestern NM, southern AZ; rare in western TX

Q. turbinella (shrub live oak): NM AZ CO UT, southern NV; rare in western TX, southwestern CA

Q. vacciniifolia (huckleberry oak): CA, western NV, southwestern OR

Q. vaseyana (Vasey's oak): central & western TX

Q. velutina (black oak): ME NH VT MA CT RI NY PA NJ DE MD VA NC SC GA AL MS LA WV KY TN MO AR OH IN IL IA MI WI, eastern MN KS TX OK, southeastern NE, northern FL

Q. viminea (Sonoran oak): rare in southeastern AZ

Q. virginiana (live oak): SC FL AL MS LA TX, eastern NC, southern GA; rare in southern OK, southeastern VA, southwestern TN

Q. welshii (shinnery oak): northern AZ, western CO, northwestern NM, southeastern UT

Q. wislizeni (interior live oak): CA

Index

A

Acer, 41

acorn moth, 11, 26

acorns

  ants and, 24–26

  as food for creatures, 11, 18

  germination/growth, 9–10

  jays and, 14–16, 21

  masting, 17–20

  shape, 120

  size, 118–120

acorn weevil, 11, 21–24, 26

acorn woodpecker, 16

Ajo Mountain oak, 173, 179

alder woolly aphid, 137

Amelanchier canadensis, 39

American chestnut, 17

American crow, 16

American elm, 10

Amorbia humerosana, 127

Amynthas, 55

Angel Oak, 124

annelids, 11

annual cicadas, 89–91, 138–141

ants, 24–26, 60, 70, 129

aphids, 88, 136

apple oak gall, 64

Argyrotaenia quercifoliana, 127

Arizona walking stick, 144

Arizona white oak, 81, 173, 179

Arkansas oak, 169, 170, 172, 174, 179

arthropods, 11, 52, 56

Asclepias, 42

Asian long-horned beetle, 84

Asian worms, 55

aspen, 54

assassin bug, 127

Atlides halesus, 103–104

Atymna, 91

autumn olive, 13, 74

azalea, 147

B

bacteria, 55
bald-faced hornet, 127, 128
Baltimore oriole, 71
barred owl, 32
bastard oak, 169, 171, 172, 175, 183
bats, 68, 124
Baumgarten, Tammany, 116
Beal, Christy, 76
bear, 11
bear oak, 167, 169, 170, 175, 181
Bedford Oak, 124
beech, 13, 30
beech nuts, 13, 14
bees, 60
Betula, 41
birch, 41, 54, 76
bird migration, 72–76, 84
birds, 11, 16, 18, 31, 68, 70, 71–76,
    78, 98–99, 110, 120, 127, 138,
    150–151
bison, 155
black bear, 16, 124
black-eyed Susan, 150
black gum, 76
blackjack oak, 168, 170, 171, 173,
    175, 176, 181
black oak, 167, 168, 169, 171, 172,
    175, 176, 184
blackpoll warbler, 71
black-throated blue warbler, 73
Blastobasis glanulella, 26
blotch leaf miner, 126

blotch mines, 125–126
blueberries, 147
blue-gray gnatcatcher, 71
bluejack oak, 169, 170, 172, 173,
    175, 181
blue jay, 14–16, 150
blue oak, 45, 177, 178, 180
bluff oak, 170, 171, 175, 179
bobcat, 124
bobwhite quail, 16
boletes, 52
Bombyx, 83
Bombyx mori, 82
bottomland post oak, 169, 171, 172,
    183
Boynton's sand post oak, 171, 179
braconids, 70
bristletails, 51
Buckley's oak, 173, 179
buds, 59–60
burningbush, 74
bur oak, 119, 167, 168, 171, 172, 174,
    175, 176, 177, 181
bush honeysuckle, 74
butterflies, 11, 42, 51, 77, 96–101

C

caddisflies, 56
California black oak, 177, 178,
    181
California hairstreak butterfly, 96
California mistletoe, 102
California scrub oak, 178, 179

Callery pear, 70, 74, 82
'Bradford', 82
Calycopis cecrops, 100
cambium, 125
Cameraria cincinnatiella, 126
Cameraria hamadryadella, 126
Canada warbler, 71
canyon live oak, 173, 177, 178, 180
canyon oak group, 11
carabids, 51
carbon dioxide, 86, 122–123
carbon sequestration, 10, 48, 122–123
Carpenter, Steve, 28
Carpinus caroliniana, 39
caterpillars, 32–34, 51, 74, 76, 77–84, 95–98, 100–101, 105, 110–118, 125–133, 145–148
catkins, 66–68, 67
Catocala, 79
cecropia moth, 68
Cedros Island oak, 178, 180
centipedes, 52
Cercis canadensis, 39
Chapman's oak, 170, 171, 180
Charadra deridens, 116–118
checkered fringed prominent, 132, 133
chemical defenses, 42–43
cherry, 41, 76
cherry-bark oak, 168, 169, 171, 172, 174, 182
chestnut oak, 167, 169, 171, 174, 176, 182

chickadee, 75–76, 150, 151
Chihuahuan oak, 173, 180
chinkapin oak, 167, 168, 170, 171, 173, 175, 176, 182
chipmunks, 16, 21
Chisos oak, 173, 179
cicada killers, 138–141
cicadas, 51, 88–91, 138–141
Cladrastis, 38
Claytonia virginica, 41
Clethra alnifolia, 103
climate change, 123, 153
coastal live oak, 177, 178, 179
coastal sage scrub oak, 178, 180
cogwheel assassin bug, 129
collembola, 51
common yellowthroat, 71
coneheads, 51
Cooper's hawk, 15
Cornus florida, 39
Corythucha arcuata, 134
cottonwood, 54
crayfish, 56
crickets, 148–150
crowned slug, 112
Curculio, 21
Cutting, Brian, 110
cynipids, 59–66

D

Dalcerides ingenita, 80–81
dalcerids, 80–81
damsel bug, 127

Danaus, 42
Darlington oak, 119, 168, 169, 170, 172, 174, 180
Datana ministra, 116, 117
datanas, 116
Davis Mountain oak, 173, 180
decomposers, 52–55, 57
deer, 10, 16, 18
deer oak, 178, 183
Diapheromera arizonensis, 144
Diapheromera femorata, 143
diplurans, 51
disease, 15–16, 153
dogwood, 39, 147
dogwood sawflies, 137
Dolbear, Amos, 148
duck, 16
Dunn's oak, 174, 178, 182
duskywing skippers, 145–147
dwarf chinkapin oak, 45, 167, 169, 170, 172, 174, 175, 176, 177, 183
dwarf live oak, 170, 172, 182
dwarf oak, 173, 181

E

early button slug, 112
earthworms, 51
eastern kingbird, 71
eastern towhee, 16
echinacea, 150
ecosystem services, 11–12, 15, 121–125, 153–156

Edwards' hairstreak butterfly, 101
emarginea, 103, 105
Emarginea percara, 103, 105
emerald ash borer, 84
Emory oak, 81, 173, 180
"enemy release", 84
energy allocation, 20
Engelmann's oak, 178, 180
English oak, 10
Epicallima argenticinctella, 126
Erynnis brizo, 145
Erynnis horatius, 145
Erynnis juvenalis, 145, 146
Erynnis propertius, 145
Erynnis zarucco, 145
Euodynerus leucomelas, 128
evening primrose, 150

F

fall cankerworm, 79
fence lizard, 11
ferns, 147
filament bearer, 96
flying squirrel, 16
food webs, 11, 36, 39, 40, 48
fringed looper, 79
fungi, 55, 124

G

gall wasp, 11, 51, 59–66
Gambel's oak, 45, 173, 176, 177, 180

Georgia oak, 170, 171, 180
golden-crowned kinglet, 31–32
goldenrod, 103, 150
gold-striped leaftier, 127
grasshoppers, 88
Graves' oak, 173, 180
gray oak, 173, 177, 180
gray squirrel, 16
great purple hairstreak butterfly,
    103–104
gregarious feeders, 116–118
gregarious oak leaf miner, 126
ground beetle, 51
ground sloth, 28, 155
growth rate, 10–11
gypsy moth, 82–84, 153

H

habitat loss, 153
hag moth, 112
hairstreak butterflies, 96–101,
    103–104
Havard's oak, 173, 176, 180
hazelnut, 147
hemipterans, 88
hemlock woolly adelgid, 84
Hercules' club, 103
hermit thrush, 71
hickory, 40, 54
Hinckley's oak, 173, 181
Horace's duskywing skipper, 145
huckleberry oak, 178, 183

I

ichneumonids, 70
inchworms, 78, 79, 84, 128–129
indigo bunting, 71
insects, 31
interior live oak, 177, 178, 184
invasive pests, 82–84, 153
invasive plants, 13, 55, 74
ironwood, 39, 76, 147
island live oak, 177, 183
Isochaetes beutenmuelleri, 110–112

J

Japanagromyza viridula, 125–126
Japanese beetle, 84
Japanese honeysuckle, 13
Japanese stiltgrass, 55
jays, 14–16, 21, 150
jewel caterpillar, 80–81
John-Tucker's oak, 181
jumping bristletail, 51
jumping spiders, 99–100
jumping worms, 55
Juvenal's duskywing skipper, 145,
    146

K

katydids, 51, 88, 106–110
Kent's geometer, 95, 96
Kentucky coffee, 76

Kentucky warbler, 71
keystone plants, 39
king's oak, 10

L

lace bugs, 51, 133–137
lace-capped caterpillar, 132, 133
lacewings, 136–137
Lacey's oak, 173, 181
Lactarius, 52
Lascoria ambigualis, 54
late-leaf oak, 174, 183
laugher caterpillar, 116–118
laurel oak, 168, 169, 172, 174, 181
leaf burning, 57
leaf litter, 11, 51–55, 100–101,
    146–148
leaf miners, 125–126
leafrollers, 127
leaf shape, 85–87
leaf skeletonizing, 127
leaf structure, 125
leather oak, 178, 180
Lewis's woodpecker, 16
limacodids, 112
Lindera benzoin, 39
Linnaeus, Carl, 79
Liquidambar styraciflua, 39
Liriodendron, 42
Liriodendron tulipifera, 42
live oak, 120, 168, 169, 171, 172,
    175, 184

Logan, William Bryant, 10
lycaenids, 100–101
Lymantria dispar, 82–84

M

machilids, 51
Machimia tentoriferella, 127
magnolia warbler, 71, 72
mammoths, 28
maple, 41, 54
maple-leaf oak, 174, 179
marcescence, 27–30, 51
masting, 17–20, 91
mastodons, 28, 155
mayflies, 56
Megatherium, 28
Mesochorus discitergus, 130
Mexican blue oak, 45, 81, 173, 182
Mexican oak, 173, 179
mice, 18
Microcentrus, 91
microclimates, 123
Microstegium vimineum, 55
milkcaps, 52
milkweed, 42
millipedes, 52
mistletoe, 102–103, 105
Mitchell, Adam, 74
mites, 52
Mohr's oak, 174, 182
mollusks, 11
monarch butterfly, 42

morels, 52
morning glory prominent, 133
Morton Arboretum, 153
moths, 11, 26, 51, 77, 80–84, 96, 100
mulch, 57–58
multiflora rose, 13, 74
mushrooms, 52
mycorrhizae, 55, 56, 123
myrtle oak, 170, 171, 182

N

Narango, Desiree, 75–76
Nason's slug, 112, 114
natural selection, 16, 65, 70, 82, 86,
    98, 108, 109, 116, 120, 127, 137
Nematocampa resistaria, 96
nematodes, 52
Neoxabea bipunctata, 148–150
net-leaf oak, 173
net-leaf white oak, 172, 183
non-native plants, 74–76
northern cardinal, 35
northern pin oak, 174, 175, 176, 180
northern red oak, 45, 46, 119, 167,
    168, 169, 171, 172, 174, 176,
    183
northern walking stick, 143
northern white oak, 45, 167, 168,
    169, 170, 172, 174, 175, 176,
    179
notodontids, 132–133
nuclear polyhedrosis viruses, 70

O

oak, 38, 76
oak beauty caterpillar, 129
oak lace bug, 134–136
oak leaf mold, 52
oak leaf scorch, 153
oak treehopper, 91–95
oak wilt, 16, 153
Oecanthus fultoni, 148
Oglethorpe oak, 169, 171, 182
Oligocentria lignicolor, 133
Oligocentria semirufescens, 133
open forests, 154
opossum, 11, 16, 124
orange-headed epicallima, 126
orchard oriole, 71
Oregon white oak, 153, 154, 177,
    178, 180
ovenbird, 71
overcup oak, 168, 170, 171, 173, 175,
    181
owls, 68

P

pachysandra, 147
Pacific oak, 178, 182
paper wasp, 127
parasitic wasps, 116, 127, 130
parasitoids, 64–65, 70, 79, 96, 127–
    128, 130
Parrhasius m-album, 96

periodical cicadas, 89–91, 138
phasmids, 142–144
Phoradendron, 102, 103
pill bug, 52
pine, 41, 76
pin oak, 167, 168, 169, 171, 172, 174, 175, 176, 182
Pinus, 41
pirate bug, 127
planthoppers, 51, 137, 138
plateau oak, 173, 180
Platycotis, 91–95
Platycotis vittata, 91–95
Podisus, 129
pollen, 66–68
pollination, 18–20
polyphemus moths, 68–70
porcelainberry, 74
porcelain gray, 79
post oak, 167, 168, 169, 171, 172, 175, 176, 183
potter wasp, 127, 128
praying mantid/mantis, 88
predators, 127–128, 138–141
Propertius duskywing skipper, 145
proturans, 51
Prunus, 41
Prunus americana, 103
Psaphida thaxteriana, 81–82
Pterophylla camellifolia, 110
purple-crested slug, 112
puss caterpillar, 112

Q

qualitative defenses, 42
quantitative defenses, 42
queen butterfly, 42
Quercus, 38, 41
Quercus acerifolia, 174, 179
Quercus agrifolia, 177, 178, 179
Quercus ajoensis, 173, 179
Quercus alba, 45, 167, 168, 169, 170, 172, 174, 175, 176, 179
Quercus arizonica, 173, 179
Quercus arkansana, 169, 170, 172, 174, 179
Quercus austrina, 170, 171, 175, 179
Quercus berberidifolia, 178
Quercus bicolor, 167, 168, 170, 171, 175, 176, 179
Quercus boyntonii, 171, 179
Quercus buckleyi, 173, 179
Quercus canbyi, 173, 179
Quercus carmenensis, 173, 179
Quercus cedrosensis, 178, 180
Quercus chapmanii, 170, 171, 180
Quercus chihuahuensis, 173, 180
Quercus chrysolepis, 173, 177, 178, 180
Quercus coccinea, 167, 168, 170, 171, 175, 176, 180
Quercus cornelius-mulleri, 178, 180
Quercus depressipes, 173, 180
Quercus douglasii, 45, 177, 178, 180

Quercus dumosa, 178, 180
Quercus durata, 178, 180
Quercus ellipsoidalis, 174, 175, 176,
 180
Quercus emoryi, 173, 180
Quercus engelmannii, 178, 180
Quercus falcata, 168, 169, 170, 172,
 174, 180
Quercus fusiformis, 173, 180
Quercus gambelii, 45, 173, 176, 177,
 180
Quercus garryana, 153, 154, 177, 178,
 180
Quercus geminata, 170, 171, 180
Quercus georgiana, 170, 171, 180
Quercus gravesii, 173, 180
Quercus grisea, 173, 177, 180
Quercus havardii, 173, 176, 180
Quercus hemisphaerica, 119, 168, 169,
 170, 172, 174, 179, 180
Quercus hinckleyi, 173, 181
Quercus hypoleucoides, 173, 181
Quercus ilicifolia, 167, 169, 170, 175,
 181
Quercus imbricaria, 168, 169, 170,
 172, 174, 175, 176, 181
Quercus incana, 169, 170, 172, 173,
 175, 181
Quercus inopina, 170, 172, 181
Quercus intricata, 173, 181
Quercus john-tuckeri, 178, 181
Quercus kelloggii, 177, 178, 181
Quercus laceyi, 173, 181

Quercus laevis, 168, 169, 171, 181
Quercus laurifolia, 168, 169, 172, 174,
 181
Quercus lobata, 177, 178, 181
Quercus lyrata, 168, 170, 171, 173,
 175, 181
Quercus macrocarpa, 119, 167, 168,
 171, 172, 174, 175, 176, 177,
 181
Quercus margarettae, 169, 170, 172,
 173, 181
Quercus marilandica, 168, 170, 171,
 173, 175, 176, 181
Quercus michauxii, 168, 170, 171,
 173, 175, 182
Quercus minima, 170, 172, 182
Quercus mohriana, 174, 182
Quercus montana, 167, 169, 171, 174,
 176, 182
Quercus muehlenbergii, 167, 168, 170,
 171, 173, 175, 176, 182
Quercus myrtifolia, 170, 171, 182
Quercus nigra, 87, 168, 169, 171, 172,
 174, 182
Quercus oblongifolia, 45, 173, 182
Quercus oglethorpensis, 169, 171, 182
Quercus pacifica, 178, 182
Quercus pagoda, 168, 169, 171, 172,
 174, 182
Quercus palmeri, 174, 178, 182
Quercus palustris, 167, 168, 169, 171,
 172, 174, 175, 176, 182
Quercus parvula, 178, 182

Quercus phellos, 45, 87, 168, 169, 171, 172, 174, 182

Quercus polymorpha, 172, 183

Quercus prinoides, 45, 167, 169, 170, 172, 174, 175, 176, 177, 183

Quercus pumila, 170, 172, 183

Quercus pungens, 173, 183

Quercus robur, 10

Quercus robusta, 173, 183

Quercus rubra, 45, 46, 119, 167, 168, 169, 171, 172, 174, 176, 183

Quercus rugosa, 173, 183

Quercus sadleriana, 178, 183

Quercus shumardii, 45, 168, 169, 171, 172, 174, 176, 183

Quercus similis, 169, 171, 172, 183

Quercus sinuata, 169, 171, 172, 175, 183

Quercus stellata, 167, 168, 169, 171, 172, 175, 176, 183

Quercus tardifolia, 174, 183

Quercus texana, 171, 172, 175, 183

Quercus tomentella, 177, 183

Quercus toumeyi, 174, 183

Quercus turbinella, 174, 177, 178, 183

Quercus vacciniifolia, 178, 183

Quercus vaseyana, 173, 183

Quercus velutina, 167, 168, 169, 171, 172, 175, 176, 184

Quercus viminea, 172, 184

Quercus virginiana, 120, 168, 169, 171, 172, 175, 184

Quercus welshii, 174, 177, 184

Quercus wislizeni, 177, 178, 184

R

rabbit, 16

raccoon, 11, 16, 124

rat snake, 11

Rectiostoma xanthobasis, 105

red-banded hairstreak butterfly, 100

red-bellied woodpecker, 16

redbud, 39

red-eyed vireo, 71

red oak group, 11, 17

red squirrel, 16

redstart, 71

red-washed prominent, 133

Richard, Melissa, 74

robust oak, 173, 183

rodents, 11

root systems, 10–11, 47, 122

rot, 124–125

roundworm, 52

running oak, 170, 172, 183

russulas, 52

S

saddleback caterpillar, 112, 115

Salix, 41

salticids, 99–100

sandhill oak, 170, 172, 181

sand live oak, 170, 171, 180
sandpaper oak, 173, 183
sand post oak, 169, 170, 172, 173, 181
Santa Cruz Island oak, 178, 182
satin moth, 84
Satyrium californica, 98
Satyrium edwardsii, 101
sawfly, 137
scarlet oak, 167, 168, 170, 171, 175, 176, 180
Schizura ipomaeae, 132
Schizura unicornis, 133
scrub oak, 178, 180
seed-cachers, 150–151
Selenia kentaria, 95, 96
serpentine mines, 126
serviceberry, 39, 40
Setophaga magnolia, 71
shingle oak, 67, 168, 169, 170, 172, 174, 175, 176, 181
shinnery oak, 174, 177, 184
shothole leaf miner, 125–126
shrub live oak, 174, 177, 178, 183
Shumard's oak, 45, 168, 169, 171, 172, 174, 176, 183
silk moth, 68, 82
silver-leaf oak, 173, 181
skiff moth, 112, 113
skippers, 96, 145–147
sleepy duskywing skipper, 145
slugs, 52, 110–112
smaller parasa, 112

Smilia, 91
snails, 52
snowy tree cricket, 148
social hornet, 139
social wasp, 127
soil erosion, 56
soil stabilization, 10
Solidago, 103
solitary oak leaf miner, 126
solitary wasp, 127
Sonoran oak, 172, 184
Sourakov, Andrei, 99–100
southern red oak, 168, 169, 170, 172, 174, 180
sow bug, 52
sphecid wasps, 90
Sphecius speciosus, 138–141
spicebush, 39, 76
spiders, 11, 31, 52, 70, 99–100, 127
spiny oak slug, 112
spraying, 79
spring beauty, 41
spring ephemerals, 41, 55, 147
springtails, 51, 52
spun glass caterpillar, 110–112
spun glass slug moth, 110–112
squirrels, 13, 16, 18, 21, 124, 138, 154
stick-mimic inchworm, 128–129
stink bug, 127, 129
stoneflies, 56
streams, 56–57
striped oak leaftier, 128

sudden oak death, 16, 153
Svendsen, Claus, 28
swamp chestnut oak, 168, 170, 171, 173, 175, 182
swamp white oak, 167, 168, 170, 171, 175, 176, 179
sweetgum, 39, 54
sweet pepperbush, 103
sycamore, 10

T

Temnothorax, 24–26
Temnothorax longispinosus, 25
tettigoniids, 107
Texas red oak, 171, 172, 175, 183
Thaxter, Roland, 81
Thaxter's sallow, 81–82
titmice, 150, 151
Toumey's oak, 174, 183
tree cricket, 51
treehoppers, 51, 88, 91–95
trillium, 55
trout lily, 55
Trouvelot, E. Leopold, 82–84
truffles, 52
Tucker's oak, 178
tufted titmice, 16
tulip tree, 40, 42, 54
turkey, 16
turkey oak, 168, 169, 171, 181
two-spotted tree cricket, 148–150

U

underwings, 79–80
unicorn caterpillar, 132, 133
United Kingdom, 153

V

valley oak, 177, 178, 181
Vasey's oak, 173, 183
vespid wasp, 70
viburnum, 40, 147
viburnum leaf beetle, 84
violet, 147
Virginia creeper, 147
vole, 10, 16

W

walking sticks, 51, 142–144
walnut woolly worm, 137
warblers, 76
wasps, 60, 64, 70, 90, 96, 116, 138, 141
water infiltration, 55–56
water oak, 87, 168, 169, 171, 172, 174, 182
watershed management, 57
wax, 137
weevils, 11, 21–24, 26
western hemlock, 49
white-breasted nuthatch, 16

white-eyed vireo, 71
white-footed mouse, 10, 16
white-lined leafroller, 127
white M hairstreak butterfly, 96
white oak, 9–11, 13, 19, 27, 48, 61
white-tailed deer, 10, 16
wild ginger, 147
wild plum, 103
willow, 41, 76
willow oak, 45, 87, 168, 169, 171, 172, 174, 182
winter moth, 84
witchhazel, 147
wood duck, 16
worms, 55

X

xylem, 89, 90, 124

Y

yellowjacket, 127, 128, 139
yellow-necked caterpillar, 116, 117
yellow-shafted flicker, 16
yellow-shouldered slug, 112
yellow-vested moth, 105
yellow warbler, 71
yellow-winged oak leafroller, 127
yellowwood, 38

Z

Zanthoxylum clava-herculis, 103
Zarucco duskywing skipper, 145